This book and your GCSE course

Awarding Body	AQA	EDEXCEL A	EDEXCEL B
Web address	www.aqa.org.uk	www.edexcel.org.uk	
Syllabus number	3411	1522	1529
Modular tests	None	None	6 tests each of 20 mins 30%
Terminal papers	1 paper 135 mins 80%	One 90 min and one 60 min paper	3 × 30 mins. 2 papers on core 1 on extension. 50%
Coursework	20%	20%	20%
Core Biology			
1. Cell activity	10.1–10.2	B1	M7
2. Humans as organisms	10.4–10.12	B1	M1, M8
3. Green plants as organisms	10.13–10.15	B2	M7
4. Variation, inheritance and selection	10.16	B3	M2, M8
5. Living things in their environment	10.20–10.23	B4	M2, M7
Extension Biology			
6. Classification			
Principles			
The five kingdoms	10.3	B5	M13
Viruses	10.3	B5	M13
7. Adaptation			
Living in water	10.25, 10.60		
Living on land	10.25, 10.61		
8. Microbes and food			
Growing microbes	10.30, 10.31	B5, B6	M13
Preserving food			
Uses of microbes	10.30, 10.33	B5, B6	M13
9. Microbes, waste and fuel			
Sewage treatment		B5	M13
Fuels	10.31		
10. Microbes and disease			
Causes of disease	10.28, 10.31	B5	M13
Spread of disease	10.28, 10.32	B5	M13
Antibiotics	10.28, 10.33	B5	M13
Vaccines	10.28, 10.34	B5	M13
11. Genetics and genetic engineering			
Structure of DNA		B6	M14
Protein synthesis		B6	M14
Mutations		B6	M14
Gene detection		B6	M14
Genetic engineering		B6	M14
12. Further physiology			
Food and feeding	10.24, 10.27, 10.30		
Excretion	10.24, 10.27, 10.31		
Co-ordination	10.24, 10.27, 10.32		
13. Food production			
World food shortage		B6	M14
Agriculture			
Plant disease and its control			
Hormones and food production		B6	M14
14. Further ecology			
Ecosystems			

Visit your awarding body for full details of your course or download your complete GCSE specifications.

Use these pages to get to know your course
- Make sure you know your exam board
- Check which specification you are doing

- Know how your course is assessed:
  - what format are the papers?
  - how is coursework assessed?
  - how many papers?

OCR A option A	OCR A option B	WJEC	NICCEA
www.ocr.org.uk		www.wjec.co.uk	www.ccea.org.uk
1980		117	
None			
90 min paper on Core, 45 min on Core and Extension A and B		FT 120 mins HT 150 mins	P1 HT 90 mins FT 60 mins P2 HT 120 mins FT 60 mins
20%	20%	20%	25%
1, 10		1	3.1
2, 3, 4, 6, 8, 9	2, 3, 4, 6, 8, 10	2	3.1
10		3	3.1
10	10	4	3.1
7		5	3.1
A2, A4	B2		
A2, A4	B2	B1	3.2
A2, A4	B2	B1	
A2			
A2			
A3	B2		3.4
A3	B2	B1, B2	3.4
A3	B2	B1, B2	3.4
	B2	B3, B4	
	B2	B3, B4	
A3, A4	B2	B2, B5	3.4
A3, A4	B2	B2, B5	3.4
A3, A4	B2	B2, B5	3.4
A3, A4	B5	B2, B5	3.4
A3	B4	B6	3.3
A3	B4	B6	3.3
A3	B4	B6	3.3
A3	B4	B6	3.3
A3	B4	B6	3.3
A1			3.1
A1		B5, B6	3.1
A1			3.1
	B3	B6	
	B3		B6
	B1		3.2

Preparing for the examination

Planning your study

The final three months before taking your GCSE examination are very important In achieving your best grade. However, the success can be assisted by an organised approach throughout the course.

- After completing a topic in school or college, go through the topic again in your Revise GCSE Biology Study Guide. Copy out the main points again on a sheet of paper or use a highlighter pen to emphasise them.
- A couple of days later, try to write out these key points from memory. Check differences between what you wrote originally and what you wrote later.
- If you have written your notes on a piece of paper, keep this for revision later.
- Try some questions in the book and check your answers.
- Decide whether you have fully mastered the topic and write down any weaknesses you think you have.

Preparing a revision programme

At least three months before the final examination, go through the list of topics in your Examination Board's specification. Go through and identify which topics you feel you need to concentrate on. It is a temptation at this time to spend valuable revision time on the things you already know and can do. It makes you feel good but it does not move you forward.

When you feel you have mastered all the topics, spend time trying past questions. Each time check your answers with the answers given. In the final couple of weeks go back to your summary sheets (or highlighting in the book).

How this book will help you

Revise GCSE Biology Study Guide will help you because:

- it contains all the essential content for your GCSE course without the extra material that will not be examined
- it contains Progress Checks and GCSE questions to help you to confirm your understanding
- it gives sample GCSE questions with answers and advice from an examiner on how to improve
- examination questions from 2003 are different from those in 2002 or 2001. Trying past questions will not help you when answering some parts of the questions in 2003. The questions in this book have been written by experienced examiners who are writing the questions for 2003 and beyond
- the summary table will give you a quick reference to the requirements for your examination
- marginal comments and highlighted key points will draw to your attention important things you might otherwise miss.

Five ways to improve your grade.

1. Read the questions carefully

Many students fail to answer the actual question set. Perhaps they misread the question or answer a similar question they have been set before. Read the question once right through and then again more slowly. Some students underline or high light key words in the question as they read it through. Questions at GCSE contain a lot of information. You should be concerned if you are not using the information in your answer.

2. Give enough detail

If a part of a question is worth three marks, you should make at least three separate points. Be careful that you do not make the same point three times. Approximately 25% of the marks on your final examination papers are awarded for questions requiring longer answers.

3. Quality of Written Communication (QWC)

From 2003 some marks on the GCSE papers are given for the quality of your written communication. This includes correct sentence structures, correct sequencing of events and use of scientific words.
Read your answer through slowly before moving on to the next part.

4. Correct use of scientific language

There is important Scientific vocabulary that you should use. Try to use the correct biological terms in your answers and spell them correctly. The way biology language is used is often a difference between successful and unsuccessful students. As you revise make a list of the biological terms that you meet and check that you understand the meaning of these words.

5. Show your working

All Biology papers include some calculations. You should always show your working in full. Then if you make an arithmetical mistake, you may still receive marks for correct biology. Check that your answer is given to the correct number of significant figures and give the correct unit.

Core material

Topic	Section	Studied in class	Revised	Practice questions
1.1 Cell structure and division	Animal and plant cells			
	Nuclei, chromosomes and genes			
	Cell division			
	Cell specialisation			
1.2 Transport in cells	Diffusion and osmosis			
	Active transport			
2.1 Nutrition	What are enzymes?			
	Digestion in the body			
	Absorption			
2.2 Circulation	Blood			
	Blood vessels			
	The circulation			
	Exchange at the tissues			
2.3 Breathing	Gaseous exchange			
2.4 Respiration	Aerobic respiration			
	Food and energy			
	Anaerobic respiration			
2.5 The nervous system	Responding to stimuli			
	Receptors			
	Neurones and responses			
2.6 Hormones	What is a hormone?			
	Hormones and reproduction			
	Using hormones			
2.7 Homeostasis	Blood glucose			
	The kidneys – control of waste			
	The kidneys – control of water			
	Temperature control			
2.8 Health	Causes of disease			
	Preventing pathogens from entering the body			
	Diseases caused by smoking			
	The action of other drugs			
3.1 Green plants as organisms	Nutrition			
	Limiting factors			
	Mineral salts			
3.2 Plant hormones	Control of plant growth			
3.3 Transport in plants	Transpiration			
3.4 Support	How water supports a plant			
3.5 Sugar transport	Phloem			
4.1 Variation	How sexual reproduction leads to variation			
	Mutation – a source of variation			
4.2 Inheritance	Sex determination			
	Monohybrid inheritance			
	Inherited diseases			
	Cloning, selective breeding and genetic engineering			
4.3 Evolution	Evidence for evolution			
5.1 Living together	Competition			
	Predators and prey			
	Adaptation			
	Cooperation			
5.2 Human impact on the environment	Population size			
	Pollution			
	Over-exploitation			
	Conservation			
5.3 Energy and nutrient transfer	Energy transfer			
	Food production			
	Decomposers			
	Nutrient cycles			

Cell structure and division

The following topics are covered in this section:

- ● *Cell structure and division*
- ● *Transport in cells*

What you should know already

Finish the passages using words from the list. You may use the words more than once.

cell membrane   **cells**   **chloroplasts**   **fertilise**   **nucleus**   **photosynthesis**   **respiration**
sensitivity   **seven**   **sperm**   **swim**   **tail**   **waste**   **specialised**

All organisms are made up of units called 1._____. These units are surrounded by a 2._____ that controls what enters and leaves. The 3._____ is the control centre of the cell. Plant cells contain 4._____ that make food by 5._____.

The diagram shows a type of animal cell.

Most cells are 6._____ for the job that they perform. The diagram above illustrates a 7._____ cell. The job of this cell is to join with or 8._____ an ovum. To help it to do this, it has a 9._____ so that it can 10._____ towards the ovum.

Living organisms are different to non-living material because they carry out 11._____ vital processes. These are often called characteristics of living organisms. The ability to respond to changes occurring around them is called 12._____ Excretion is the ability to remove 13._____ products that have been produced by the organism. The release of energy from food molecules is called 14._____.

ANSWERS

1. cells; 2. cell membrane; 3. nucleus; 4. chloroplasts; 5. photosynthesis; 6. specialised; 7. sperm; 8. fertilise; 9. tail; 10. swim; 11. seven; 12. sensitivity; 13 waste; 14. respiration

1.1 Cell structure and division

After studying this section you should be able to:

LEARNING SUMMARY

● describe the main differences between plant and animal cells
● state that the nucleus contains chromosomes
● explain why cells can divide in two different ways
● explain how certain cells are specialised for the jobs that they do.

Animal and plant cells

AQA
Edexcel A   Edexcel B
OCR A ^A^   OCR A ^B^
NICCEA
WJEC

Although plants and animals have many things in common, there are four main differences:

● plant cells have a strong cell wall made of cellulose, animal cells do not

● plant cells have a large permanent vacuole containing cell sap, vacuoles in animal cells are small and temporary

Remember that some plant cells, such as root cells, do not have chloroplasts.

● plant cells may contain chloroplasts containing chlorophyll for photosynthesis. Animal cells never contain chloroplasts

● animal cells store energy as granules of glycogen but plants store starch.

Plant and animal cells have many smaller structures in the cytoplasm. These can be seen by using an electron microscope.

Common mistake: many candidates think that only animals respire. Plants also respire and so plant cells also have mitochondria.

 KEY POINT Mitochondria **are examples of these structures and are the site of respiration in the cell.**

Fig. 1.1

Nuclei, chromosomes and genes

AQA
Edexcel A   Edexcel B
OCR A ^A^   OCR A ^B^
NICCEA
WJEC

Most cells contain a nucleus that controls all of the chemical reactions that go on in the cell. Nuclei can do this because they contain the genetic material. Genetic material controls the characteristics of an organism and is passed on from one generation to the next. The genetic material is made up of structures called chromosomes. They are made up of a chemical called **Deoxyribonucleic Acid** or **DNA**. The DNA controls the cell by coding for the making of proteins, such as enzymes. The enzymes will control all the chemical reactions taking place in the cell.

There is much more information about genes and how they work in section 4.2.

 **KEY POINT** A **gene** is a part of a chromosome that codes for one particular protein.

By controlling cells, genes therefore control all the characteristics of an organism. Different organisms have different numbers of genes and different numbers of chromosomes. In most organisms that reproduce by sexual reproduction, the chromosomes can be arranged in pairs. This is because one of each pair comes from each parent.

Cell division

AQA
Edexcel A   Edexcel B
OCR A ᴬ   OCR A ᴮ
NICCEA
WJEC

There seems to be a limit to how large one cell can become. If organisms are to grow, cells must split or divide. Cells also need to divide to make special sex cells called **gametes** for reproduction.

> **KEY POINT**
> Cells therefore need to divide for two main reasons – for growth or reproduction.

> Because the two words meiosis and mitosis are very similar, you need to spell them correctly in order to score marks in exams.

There are two types of cell division, one for each of these two reasons:

| **Mitosis** is used for growth | → | **New cells** | ← | **Meiosis** is used for gametes |

Fig. 1.2

Both of these two types of cell division have certain things in common. The DNA of the chromosomes has to be copied first to make new chromosomes. The chromosomes are then organised into new nuclei and the cytoplasm then divides into new cells.

In mitosis two cells are produced from one. As long as the chromosomes have been copied correctly, each new cell will have the same number of chromosomes and the same information.

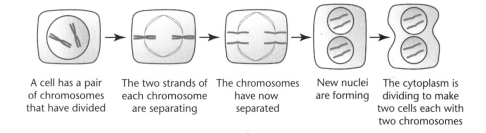

A cell has a pair of chromosomes that have divided → The two strands of each chromosome are separating → The chromosomes have now separated → New nuclei are forming → The cytoplasm is dividing to make two cells each with two chromosomes

Fig. 1.3

In meiosis, the chromosomes are also copied once but the cell divides twice. This makes four cells each with half the number of chromosomes, one from each pair.

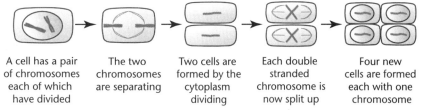

A cell has a pair of chromosomes each of which have divided → The two chromosomes are separating → Two cells are formed by the cytoplasm dividing → Each double stranded chromosome is now split up → Four new cells are formed each with one chromosome

Fig. 1.4

Cell specialisation

AQA

Edexcel A   Edexcel B

OCR A ᴬ   OCR A ᴮ

NICCEA

WJEC

By the process of mitosis a large number of cells can be produced. This enables organisms to grow or repair damaged tissue. The different cells all contain the same genes but develop differently.

KEY POINT

Cells become adapted for different functions. This is called specialisation.

Specialisation allows cells to become more efficient at carrying out their jobs.

The disadvantage of being specialised is that the cells lose the ability to take over the jobs of other cells if they are lost.

An example of a specialised cell is a nerve cell or neurone:

Fig. 1.5

There is much more about nerves and how they work in section 2.5.

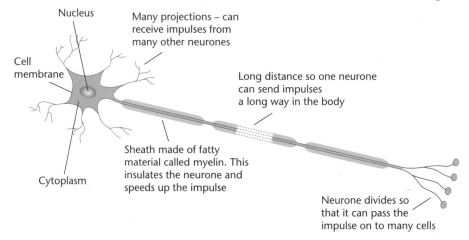

Nucleus

Many projections – can receive impulses from many other neurones

Cell membrane

Long distance so one neurone can send impulses a long way in the body

Sheath made of fatty material called myelin. This insulates the neurone and speeds up the impulse

Cytoplasm

Neurone divides so that it can pass the impulse on to many cells

- similar cells that do similar jobs are gathered together into tissues

Be careful: bone and muscle are tissues, but a bone or a muscle is an organ.

- more complicated organisms have organs that are made up of a number of tissues

- groups of organs work together in systems to carry out certain functions.

Fig. 1.6

Cells e.g. nerve cells → Tissues e.g. nerve tissues → Organs e.g. brain → Systems e.g. nervous system →

PROGRESS CHECK

1. What are cell walls made of?
2. What is the main difference between vacuoles in plant cells and those in animal cells?
3. What do mitochondria do?
4. Place these structures in order of size, largest first:
   nucleus   mitochondrion   chloroplast   liver cell   chromosome
5. Which type of cell division is used for growth?
6. How many cells are produced when one cell divides by meiosis?

1. Cellulose;   2. Vacuoles in plant cells are larger and permanent;   3. Carry out respiration;
4. Liver cell, nucleus, chloroplast, mitochondrion, chromosome;   5. Mitosis;   6. Four.

1.2 Transport in cells

LEARNING SUMMARY

After studying this section you should be able to:

- *understand how substances pass in and out of cells including:*
  - *passively by diffusion*
  - *by osmosis, which is a special type of diffusion*
  - *by active transport, which requires energy.*

Diffusion and osmosis

 AQA
 Edexcel A   Edexcel B
 OCR A ᴬ   OCR A ᴮ
NICCEA
WJEC

Definition

Diffusion is the net movement of a substance from an area of high concentration to an area of low concentration. This is down a diffusion gradient.

How does it work?

Diffusion works because particles are always moving about in a random way. This means that the particles will spread out evenly after a while.

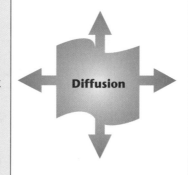

Diffusion

How fast does it work?

The rate of diffusion depends on how fast the particles move. The warmer it is, the faster they move. Smaller particles also move faster.

Examples

Oxygen diffuses into the red blood cells in the lungs and carbon dioxide diffuses out of the blood. Carbon dioxide enters leaves and leaf cells by diffusion.

Fig. 1.7

Osmosis and diffusion are called passive. This means that they do not need energy from respiration to occur. The energy comes from the movement of the particles.

Osmosis is really a special kind of diffusion. It involves the movement of water molecules. It needs a:

- **selectively permeable membrane** – the cell membrane is selectively permeable because it lets certain molecules through and not others. The water can pass through but the dissolved substance cannot

- **different concentration of solution on each side of the membrane** – water will move from the weak solution (high concentration of water) to the strong solution (low concentration of water).

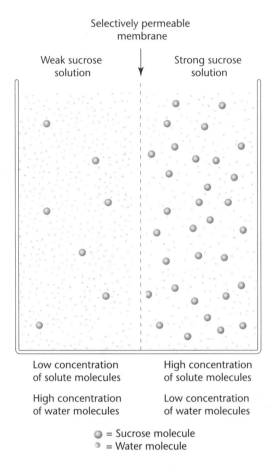

Selectively permeable membrane

Weak sucrose solution

Strong sucrose solution

Low concentration of solute molecules

High concentration of solute molecules

High concentration of water molecules

Low concentration of water molecules

= Sucrose molecule
= Water molecule

Fig. 1.8

> **KEY POINT** Osmosis is the net movement of water from a dilute to a concentrated solution through a selectively permeable membrane.

Experiments with osmosis

When plant cells gain water by osmosis, they swell. The cell wall stops them from bursting. Osmosis can be studied by placing pieces of plant tissue into different concentrations of sugar solution. If the pieces of tissue increase in mass then water has entered the tissue by osmosis. This is because the solution is weaker than the concentration inside the cells. If the tissue loses mass then water has left the tissue. By finding the point at which there is no change in mass, the concentration inside the cells can be estimated.

Turgid plant cells are very important for helping to support plants. Osmosis is also important in the uptake of water by roots. Both these are described in greater detail in sections 3.3 and 3.4.

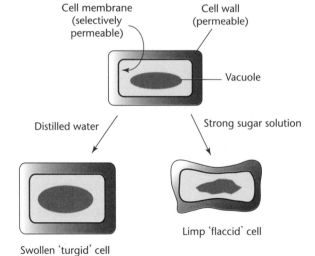

Cell membrane (selectively permeable)

Cell wall (permeable)

Vacuole

Distilled water

Strong sugar solution

Swollen 'turgid' cell

Limp 'flaccid' cell

Fig. 1.9

Active transport

AQA
Edexcel A   Edexcel B
OCR A ᴬ   OCR A ᴮ
NICCEA
WJEC

Sometimes substances have to be moved from a place where they are in low concentration to where they are in high concentration. This is in the opposite direction to diffusion and is called active transport.

Definition

Active transport is the movement of a substance against a diffusion gradient with the use of energy from respiration.

How does it work?

Proteins in the cell membrane pick up the substance and carry it across the membrane. This requires energy, which is produced in the cell from respiration.

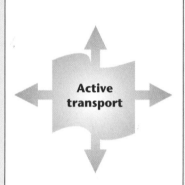

Active transport

How fast does it work?

Anything that slows down respiration will slow down the rate of active transport. This could be a poison, such as cyanide, or lack of oxygen.

Examples

Glucose is absorbed from the food into the cells of the small intestine by active transport. Minerals are absorbed into plant roots from the soil against a concentration gradient.

Fig. 1.10

PROGRESS CHECK

1. Where does the energy for diffusion come from?
2. What is a diffusion gradient?
3. Why do vegetables swell up when they are placed in a saucepan of water prior to cooking?
4. A person in a room is wearing strong scent. Why can people smell this scent more quickly on a warm day?
5. How is active transport different to diffusion?
6. Plant roots take up minerals very slowly from waterlogged soil. Why is this?

1. The movement of the particles (kinetic energy);   2. This is when a substance is not spread out evenly and is in high concentration in one area and in low concentration in an adjacent area;   3. The vegetables take in water by osmosis because their cell contents are more concentrated than the water.   4. The particles have more energy and move quicker;   5. Active transport needs energy from respiration, diffusion does not. Active transport is against a diffusion gradient but diffusion is down a diffusion gradient;   6. There is less oxygen in a waterlogged soil so respiration is slower, releasing less energy for active transport.

Sample GCSE question

1. A pupil wanted to investigate osmosis in potato tissue. He cut five cylinders from a potato and measured the mass of each. He then placed each block in a different concentration of sucrose solution.

The table shows his results:

Concentration of solution in mol per dm³	Mass of potato cylinder before soaking (in grams)	Mass of potato cylinder after soaking (in grams)	% change in mass
0.0	4.90	5.51	12.40
0.2	4.70	5.10	8.50
0.4	4.80	4.85	1.00
0.6	4.80	4.66	
0.8	5.20	4.81	−7.50

(a) Work out the percentage change in mass for the potato cylinder in the 0.6 mol per dm³ solution. **[2]**

4.66 − 4.8 = 0.14 ✓ *× 100 = −2.9* ✓

> For percentage change calculations take the starting number away from the final number and multiply by 100.

(b) Plot the results on the grid. **[3]**

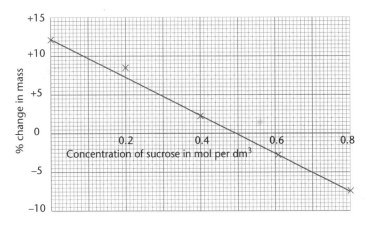

> Choose a suitable scale that will use more than half of the graph paper. Make sure that you have a long ruler in order to draw a single straight line.

(c) Finish the graph by drawing the best straight line. **[1]**

(d) Explain what happens to the potato cylinders that were placed in the sucrose solutions with a concentration of more than 0.5 mol per dm³. **[3]**

The potato blocks decreased in mass ✓*. This is because their cells are less concentrated than the sucrose solution and so lose water* ✓ *by osmosis* ✓*.*

> Do not say that the sucrose solution moved out of the potato. This is a common mistake.

Exam practice questions

1. The diagram shows a cell from a plant.

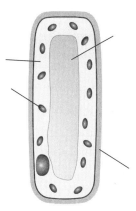

(a) **(i)** Finish the diagram by completing the labels. **[4]**

(ii) Where in a plant does this cell come from? **[1]**

(iii) Name two structures that would not be present if this was an animal cell. **[2]**

(b) This diagram has been drawn using a light microscope. Name one structure found in cells that is too small to be seen with the light microscope. **[1]**

2. The diagram shows an experiment to investigate the uptake of mineral ions into the roots of plants.

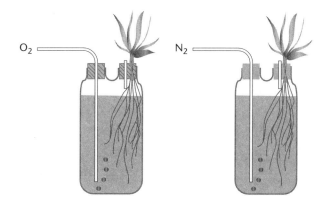

Oxygen or nitrogen gas was bubbled through the water and the uptake of minerals from the solution was measured.

(a) The plant took up more minerals when oxygen was bubbled through the solution. Explain why this is. **[3]**

(b) A waterlogged soil contains little air. Explain why farmers try to make sure their fields are well-drained. **[3]**

Humans as organisms

The following topics are covered in this section:

- Nutrition
- Circulation
- Breathing
- Respiration
- The nervous system
- Hormones
- Homeostasis
- Health

What you should know already

Finish the passages using words from the list. You may use the words more than once.

amnion	amniotic fluid	bronchi	bronchioles	carbon dioxide	cervix	digested
egestion	enzymes	fetus	fish	growth	iron	placenta
respiration	spinach	trachea	umbilical cord	uterus		

A balanced diet contains seven groups of substances. Proteins are necessary for 1._____ and for making molecules called 2._____ that speed up the rate of chemical reactions in the body. Foods such as 3._____ are a good source. Minerals are another group of substances. An example is 4._____, that is needed to make red blood cells. This is found in foods, such as 5._____. Proteins are too large to be able to pass into our bloodstream and so need to be 6._____ first. The removal of undigested food from the body is called 7._____.

The diagram shows a fetus inside a female.

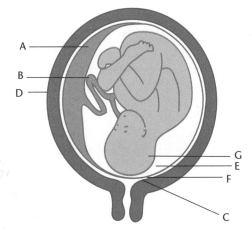

Label the structures A to G:

A= 8._____

B= 9._____

C= 10._____

D= 11._____

E= 12._____

F= 13._____

G= 14._____.

The lungs are connected to the mouth by a tube called the 15._____. This divides into two 16._____, one passing to each lung. Finer divisions of these tubes are called 17._____ and they end in millions of air sacs. The job of the lungs is to obtain enough oxygen for the process of 18._____ and to remove 19._____.

2.1 Nutrition

LEARNING SUMMARY

After studying this section you should be able to:

● *describe how food is broken down in the digestive system by enzymes*
● *locate the parts of the digestive system that produce these enzymes*
● *describe the process of absorption.*

What are enzymes?

AQA
Edexcel A   Edexcel B
OCR A ᴬ   OCR A ᴮ
NICCEA
WJEC

KEY POINT Enzymes are biological catalysts. They are produced in all living organisms and control all the chemical reactions that occur.

Most of the chemical reactions that occur in living organisms would occur too slowly without enzymes. Increased temperatures would speed up the reactions but using enzymes means that the reactions are fast enough at 37°C.

Enzymes are protein molecules that have a particular shape. They have a slot or a groove into which the substrate fits. The reaction then takes place and the products leave the enzyme.

Remember: that enzymes are present in all cells not just in the digestive system.

The substrate in a reaction is the chemical that reacts and the product is the chemical that is made.

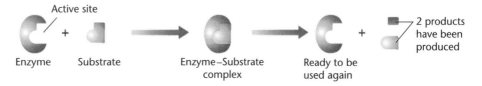

Fig. 2.1

KEY POINT This explains why enzymes are specific. Each enzyme is designed to fit only one substrate.

In the digestive system three of the main substances that need digesting are starch, proteins and fats. They are each broken down with the help of a different type of enzyme.

There are different types of amylases, proteases and lipases produced in different parts of the gut.

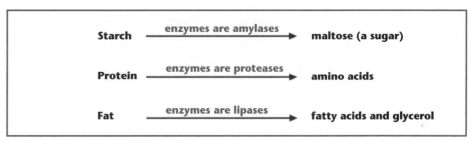

Fig. 2.2

Studying enzymes

AQA
Edexcel A   Edexcel B
OCR A ᴬ   OCR A ᴮ
NICCEA
WJEC

In order to see if starch, protein or fats have been digested, we can use **food tests**.

> These food tests are often used to study enzyme reactions.

food molecule	substance used for test	details of test	sign of a positive result
starch	iodine solution	drop iodine solution into the solution to be tested	solution turns blue-black
simple sugars	Benedict's solution	add Benedict's solution to the solution and boil in water bath for two minutes	solution turns orange-red
fats	ethanol	ethanol is shaken with the substance to be tested and then a few drops of the ethanol are dropped into water	a milky white emulsion forms in the water
protein	sodium hydroxide and copper sulphate (Biuret test)	add several drops of dilute sodium hydroxide solution followed by several drops of copper sulphate solution	solution turns purple

> The best conditions are called the optimum.

These food tests can be used to see in which conditions enzymes work best. For example the effect of temperature can be investigated.

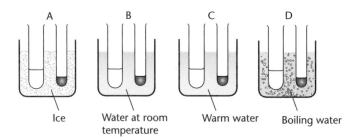

Ice   Water at room temperature   Warm water   Boiling water   **Fig. 2.3**

The starch is mixed with amylase and then small amounts are tested with iodine solution to see how fast the starch is digested. If the temperatures of the water baths are measured then a graph can be plotted to show how fast the reaction occurs at different temperatures.

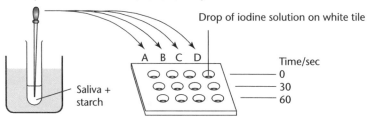

One drop from each of tubes A, B, C, D, every 30 seconds

Drop of iodine solution on white tile

A B C D

Time/sec
0
30
60

Saliva + starch

Fig. 2.4

> Enzymes don't 'die' at high temperatures, they are not living organisms.

KEY POINT

As the temperature is increased, the reaction occurs faster but above about 37°C it slows down. This is because at high temperatures the enzymes change shape or denature.

Digestion in the body

AQA
Edexcel A   Edexcel B
OCR A^A   OCR A^B
NICCEA
WJEC

As the food passes down the digestive system, different secretions are added to the food in order to digest the large molecules.

Saliva also contains mucus to lubricate the food.

Food is moved down the digestive system by muscular contractions called peristalsis.

Bile salts are not enzymes. They make the surface area of the fat droplets larger so lipase works faster. This is called emulsifying.

Secretions from the pancreas and liver are alkaline. This helps to neutralise the acid from the stomach.

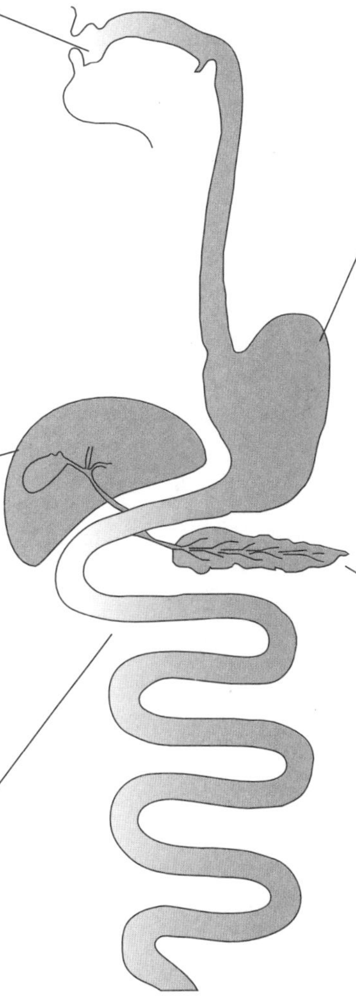

Saliva is released into the mouth from the salivary glands. It contains amylase to digest starch to maltose.

The stomach makes gastric juice, containing protease and hydrochloric acid. The acid kills microbes and creates the best pH for the protease to digest proteins.

The liver makes bile that contains bile salts. They break the large fat droplets down into smaller droplets. Bile is stored in the gall bladder.

The pancreas makes more protease and amylase. It also makes lipase to digest the fats to fatty acids and glycerol.

The small intestine makes enzymes such as maltase. This digests maltose to glucose.

Fig. 2.5

Absorption

AQA
Edexcel A   Edexcel B
OCR A^A   OCR A^B
NICCEA
WJEC

KEY POINT Simple sugars, amino acids, fatty acids and glycerol are all small enough to pass through the lining of the intestine into the blood stream. This is called absorption.

In order for the body to use food substances, they must get into the blood stream.

The second part of the small intestine is called the **ileum**. This is where **absorption** takes place. The ileum is specially adapted so that absorption can be speeded up. The surface area is increased because:

- the ileum is very long, about 5 metres in man
- the inside of the ileum is folded
- the folds have thousands of finger-like projections called **villi**
- the cells on the villi have projections called **microvilli**.

These adaptations increase the surface area by up to 600 times.

Lacteals empty their contents into the bloodstream near the heart.

The **villi** contain large numbers of capillaries to take up the products of digestion. There are also other vessels called **lacteals** that mainly absorb the products of fat digestion.

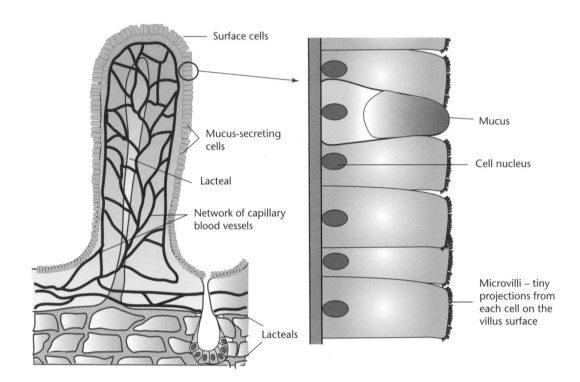

Surface cells

Mucus-secreting cells

Lacteal

Network of capillary blood vessels

Lacteals

Mucus

Cell nucleus

Microvilli – tiny projections from each cell on the villus surface

Fig. 2.6

PROGRESS CHECK

1. Name the enzyme that digests proteins.
2. Explain why lipase does not digest starch.
3. Why does boiling an enzyme prevent it from working?
4. What is the function of bile?
5. Why does the stomach produce acid?
6. What is a villus and what does it do?

1. Protease;   2. Enzymes are specific. The starch would not fit into the active site;
3. Enzymes are denatured by high temperatures;   4. Bile contains bile salts which emulsify fats;
5. The acid provides the best pH for the protease to work and kills microbes;   6. A villus is a finger-like projection from the wall of the small intestine, that increases surface area for absorption of food.

2.2 Circulation

LEARNING SUMMARY

After studying this section you should be able to:

- describe the structure and functions of blood
- describe the blood vessels that carry blood around the body
- explain how the heart circulates blood and how substances are exchanged at the tissues.

Blood

AQA
Edexcel A   Edexcel B
OCR A ᴬ   OCR A ᴮ
NICCEA
WJEC

KEY POINT

Blood consists of a straw-coloured liquid called plasma in which are suspended white blood cells, red blood cells and platelets.

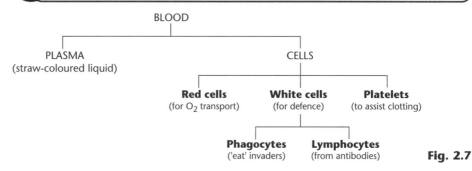

```
                        BLOOD
                          |
          ┌───────────────┴───────────────┐
       PLASMA                            CELLS
(straw-coloured liquid)                   |
                        ┌─────────────────┼─────────────────┐
                    Red cells         White cells        Platelets
                  (for O₂ transport)   (for defence)   (to assist clotting)
                                          |
                                ┌─────────┴─────────┐
                            Phagocytes         Lymphocytes
                           ('eat' invaders)   (from antibodies)
```

Fig. 2.7

The plasma and cells in blood can be separated by spinning blood in a machine called a centrifuge.

The **plasma** is about 90% water but it has a number of other chemicals dissolved in it:

- blood proteins, including some that work together with the platelets to make the blood clot

- food substances, such as glucose and amino acids

- hormones

- waste materials, such as urea

- mineral salts, such as hydrogen carbonate, the main method of carrying CO_2.

KEY POINT

Red blood cells are biconcave discs with no nucleus. They contain **haemoglobin** which carries oxygen around the body.

The structure of red blood cells makes them adapted for their job of picking up and carrying oxygen

No nucleus so more haemoglobin can fit in

White blood cells: phagocytes

Cytoplasm with large amount of haemoglobin

Shape gives a large surface area to pass oxygen through

White blood cells: lymphocytes

Fig. 2.8

For more on white blood cells see section 2.8.

KEY POINT

There are two main types of white blood cell:
- **phagocytes** engulf foreign cells, such as bacteria
- **lymphocytes** make proteins called antibodies that kill invading cells.

Blood vessels

AQA

Edexcel A   Edexcel B

OCR A ^A^   OCR A ^B^

NICCEA

WJEC

Until the 17th century, scientists had little idea how blood flowed around the body. In 1628 William Harvey published the results of his studies on the circulation of blood. He was the first person to work out the jobs of the three different types of blood vessels.

> **Harvey could not see capillaries but predicted that they must exist.**

KEY POINT

Arteries always carry blood away from the heart and **veins** carry blood back to the heart. **Capillaries** join the arteries to the veins.

The three types of blood vessel are quite different in terms of their structure because they are adapted to do different jobs:

> **Remember: arteries = A for away from the heart.**

Arteries	**Capillaries**	**Veins**
The blood is being carried away from the heart and so the pressure is high	The pressure is lower than in arteries but is still high enough to make the plasma squeeze out into the tissues	The blood returning to the heart is under low pressure
Thick wall with plenty of elastic and muscle tissue	Wall is one cell thick so that plasma can leak out	Wall is thinner than in arteries

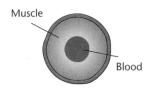

Muscle

Blood

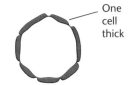

One cell thick

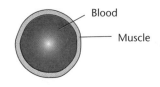

Blood

Muscle

> **The exceptions to this rule are the arteries and veins that pass to and from the lungs.**

Blood is usually oxygenated	The site of oxygen exchange with the tissues	Blood is usually deoxygenated
Valves are not needed	No valves	Valves are present to stop back-flow of blood as the pressure is low

Fig. 2.9

Valve closed

Blood flow

The circulation

AQA
Edexcel A Edexcel B
OCR A ^A^ OCR A ^B^
NICCEA
WJEC

The blood is circulated around the body by the **heart**. The heart is a muscular pump made of a special type of tissue called cardiac muscle.

KEY POINT The circulation in mammals is called a **double circulation**. This is because the blood is sent from the heart to the lungs to be **oxygenated**, but then returns to the heart to be pumped to the body.

The big advantage of a double circulation is that the blood is returned to the heart to gain high enough pressure to get through the capillaries of the body.

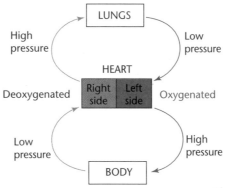

Fig. 2.10

Having a **double circulation** means that the heart has to deal with both deoxygenated and oxygenated blood at the same time. This means that the blood in the left and right side of the heart must not mix.

KEY POINT The right side of the heart carries deoxygenated blood that has returned from the body and is pumped to the lungs. The left side of the heart carries oxygenated blood.

This is an important diagram to learn. Remember that the right side of the heart is on the left as you look at it.

Notice that the wall of the left ventricle is thicker than the right ventricle. This is because it has to pump the blood further.

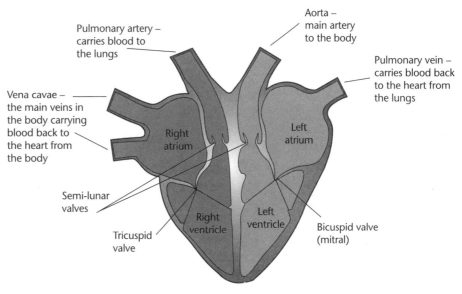

Fig. 2.11

The top two chambers are the **atria** – they receive blood from the veins.

The atria then pump the blood down into the bottom two chambers called the **ventricles** – they pump blood out to the arteries.

The various valves in the heart make sure that the blood flows in this direction and cannot flow backwards.

Exchange at the tissues

When the blood flows through the capillaries, the thin walls of the capillaries allow different substances to leave or enter the blood. The direction of movement of these substances is different in different parts of the body:

- capillaries in the lungs

In the lungs CO_2 diffuses out of blood and into the air sacs and O_2 diffuses into the blood.

> Tissue fluid does not form in the lungs. If it did it would stop gaseous exchange.

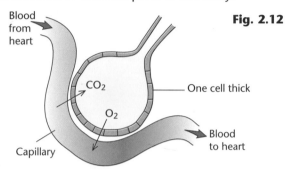

Blood from heart

CO_2

O_2

One cell thick

Capillary

Blood to heart

Fig. 2.12

- capillaries in the other tissues of the body.

The high pressure of the blood causes some of the plasma to be squeezed out of the capillaries. This is called **tissue fluid** and it carries glucose, amino acids and other useful substances to the cells.

In the rest of the tissues of the body, CO_2 diffuses into the capillaries and O_2 diffuses out of the blood into the tissues.

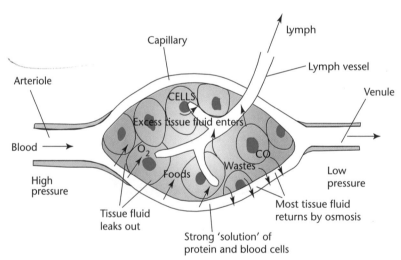

Capillary

Lymph

Lymph vessel

Venule

Arteriole

CELLS

Excess tissue fluid enters

O_2

CO_2

Foods

Wastes

Blood →

High pressure

Low pressure

Most tissue fluid returns by osmosis

Tissue fluid leaks out

Strong 'solution' of protein and blood cells

Fig. 2.13

PROGRESS CHECK

1. What substances are dissolved in plasma?
2. Why are red blood cells shaped like a biconcave disc?
3. What is the definition of an artery?
4. Why are the walls of capillaries only one cell thick?
5. What is the function of the valves in the heart?
6. What is tissue fluid?

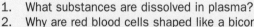

1. Proteins, dissolved food substances, minerals, waste substances, hormones; 2. To provide a larger surface area so that they can exchange oxygen faster; 3. A blood vessel that carries blood away from the heart; 4. This allows gases to diffuse across and some of the plasma to be squeezed out; 5. They stop the blood flowing backwards, i.e. back into the atria from the ventricles or back into the ventricles from the arteries; 6. This is the part of plasma that is squeezed out of the capillaries at the tissues in order to supply the cells with useful substances.

2.3 Breathing

LEARNING SUMMARY

After studying this section you should be able to:

- explain how air is drawn into and forced out of the lungs
- state the composition of the inhaled and exhaled air
- explain how the lungs are adapted for gaseous exchange.

Exchanging air

AQA
Edexcel A   Edexcel B
OCR A ^A   OCR A ^B
NICCEA
WJEC

KEY POINT Breathing is a set of muscular movements that draw air in and out of the lungs. It means that more oxygen is available in the lungs and more carbon dioxide can be removed.

> Breathing is needed in large or very active animals because they need more oxygen.

Drawing air in and out of the lungs involves changes in pressure and volume in the chest. These changes work because the **pleural** membranes form an airtight **pleural cavity**.

KEY POINT The pleural cavity is air-tight and so an increase in volume in the cavity will decrease the pressure.

> Common mistake: the lungs do not force the ribs outwards. When the ribs move this causes the lungs to inflate.

Breathing in (inhaling):

1. The intercostal muscles contract moving the ribs upwards and outwards.

2. The diaphragm contracts and flattens.

3. Both of these actions will increase the volume in the pleural cavity and so decrease the pressure.

4. Air is therefore drawn into the lungs.

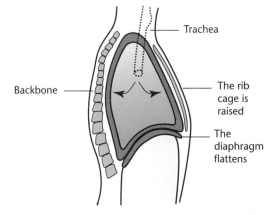

Trachea

Backbone

The rib cage is raised

The diaphragm flattens

Fig. 2.14

Breathing out (exhaling):

1. The intercostal muscles relax and the ribs move down and inwards.

2. The diaphragm relaxes and domes upwards.

3. The volume in the pleural cavity is decreased so the pressure is increased.

4. Air is forced out of the lungs.

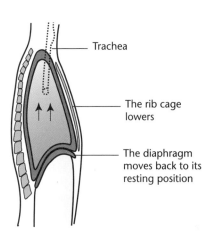

Trachea

The rib cage lowers

The diaphragm moves back to its resting position

Fig. 2.15

Whilst the air is in the lungs, the proportions of oxygen and carbon dioxide are changed.

There is still 17% oxygen in air that is breathed out. This makes artificial respiration possible.

Gas	Percentage of the gas present in	
	air breathed in	air breathed out
carbon dioxide	0.04	4
oxygen	21	17
nitrogen	78	78

Gaseous exchange

AQA
Edexcel A   Edexcel B
OCR A ^A   OCR A ^B
NICCEA
WJEC

KEY POINT Gaseous exchange occurs when oxygen diffuses from the air into the bloodstream and carbon dioxide diffuses the other way.

Gaseous exchange occurs in the millions of air sacs in the lungs. These are called **alveoli**. The structure of these alveoli makes them very efficient (well adapted) for gaseous exchange.

Remember: deoxygenated blood is not really blue!

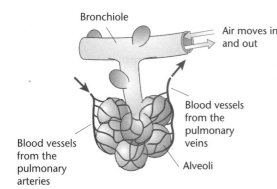

Fig. 2.16

Bronchiole

Air moves in and out

Blood vessels from the pulmonary veins

Blood vessels from the pulmonary arteries

Alveoli

Blood with a low oxygen concentration and a high carbon dioxide concentration

Air moves in and out

Blood with high oxygen concentration and a low carbon dioxide concentration

Carbon dioxide diffuses out

Oxygen diffuses in

Carbon dioxide diffuses out of blood
Oxygen diffuses into blood

Gases dissolve in layer of moisture

Wall of capillary – only one cell thick

Wall of alveolus – only one cell thick

The leaves of plants are also adapted for gaseous exchange. This is described in Chapter 3.

The adaptations are:

● the millions of alveoli provide a surface area of about 90 m²

● the many blood vessels provide a rich blood supply

● the alveoli have a thin film of moisture, so that the gases can dissolve

● the blood and air are separated by only two layers of cells.

PROGRESS CHECK

1. Why does an elephant need to breathe whilst a tree, of the same size, does not?
2. Which two structures contract to make a person breathe in?
3. What is the difference between the oxygen content of air that is breathed in compared to air that is breathed out?
4. Where does blood go to after it has been through the lungs and how does it get there?

1. A elephant is more active and so needs more oxygen; 2. The intercostal muscles and the diaphragm; 3. 21% – 17% = 4%; 4. It goes back to the left atrium of the heart in the pulmonary vein.

2.4 Respiration

After studying this section you will be able to:

- recall that respiration occurs in all living things all of the time
- understand that aerobic respiration uses oxygen and is a similar process to burning
- understand that anaerobic respiration does not use oxygen and produces different products
- understand that aerobic respiration releases more energy than anaerobic respiration
- understand what is meant by the term 'oxygen debt'.

Aerobic respiration

AQA
Edexcel A   Edexcel B
OCR A ᴬ   OCR A ᴮ
NICCEA
WJEC

Aerobic respiration is when glucose reacts with oxygen to release energy. Carbon dioxide and water are released as waste products.

 KEY POINT The equation for respiration, is the equation for photosynthesis backwards.

This process is similar to burning, but much slower.

glucose + oxygen → carbon dioxide + water + **energy**

$$C_6H_{12}O_6 + 6O_2 \rightarrow 6CO_2 + 6H_2O + $$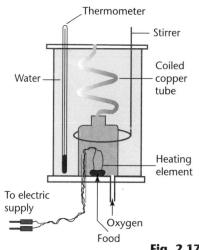

Both animals and plants respire all of the time. The rate of respiration can be estimated by measuring how much oxygen is used. The heat given off maintains our high body temperature.

Food and energy

AQA
Edexcel A   Edexcel B
OCR A ᴬ   OCR A ᴮ
NICCEA
WJEC

Different food contains different amounts of energy. Fat contains about twice the amount of energy per gram, as glucose.

Calorimeters can measure the energy content of food.

1. The food is burnt in a closed container.

2. The container is surrounded by water.

3. The increase in temperature of the water is measured.

4. The amount of energy in the food can then be calculated.

The energy used to be measured in calories. It is now measured in **joules**.

Fig. 2.17

Anaerobic respiration

AQA
Edexcel A Edexcel B
OCR A ᴬ OCR A ᴮ
NICCEA
WJEC

When not enough oxygen is available, glucose can be broken down by **anaerobic respiration**.

In humans: glucose → lactic acid + **energy**.

In yeast: glucose → carbon dioxide + ethanol + **energy**.

> You should be able to compare these anaerobic equations with the equation for aerobic respiration.

Oxygen debt

Being able to respire without oxygen sounds a great idea.

However, there are two problems:

- Anaerobic respiration releases less than half the energy of that released by aerobic respiration.
- Anaerobic respiration produces lactic acid. Lactic acid causes muscle fatigue.

What causes the oxygen debt?

When vigorous exercise takes place:

1. The muscles respire aerobically to release energy.

2. Soon the muscles require more oxygen than can be supplied by the lungs.

3. The muscles now have to break down glucose without oxygen, using anaerobic respiration.

4. Lactic acid builds up in the muscles.

5. When the vigorous exercise stops, the lactic acid is still there, and has to be broken down.

6. This requires oxygen and this 'debt' now has to be repaid.

7. Once we have breathed in enough oxygen to break down the lactic acid, the debt has been repaid.

> Have you ever wondered why once you have stopped running and using lots of energy, you are still out of breath?

KEY POINT The fitter we are, the quicker we can breath in the oxygen, and the sooner we repay the debt.

---

1. Write down the equation for aerobic respiration.
2. Write down the equation for anaerobic respiration in humans.
3. State three differences between aerobic and anaerobic respiration.
4. State what instrument can be used to measure the energy content of food.
5. State what is meant by the term 'oxygen debt'?
6. State why fit athletes can repay their oxygen debt more quickly than an unfit person.

PROGRESS CHECK

1. $C_6H_{12}O_6 + 6O_2 \rightarrow 6CO_2 + 6H_2O$ + energy; 2. Glucose → lactic acid + energy; 3. Aerobic – uses oxygen, more efficient, does not produce lactic acid. 4. Calorimeter; 5. Anaerobic respiration produces lactic acid. This has to be broken down by oxygen when exercise stops. The oxygen that is breathed in after exercise repays this debt; 6. Athletes usually have bigger more powerful lungs that can absorb oxygen faster than the lungs of an unfit person.

2.5 The nervous system

After studying this section you should be able to:

- *explain how organisms respond to stimuli*
- *explain how the eye works*
- *describe the nerve pathway of a reflex.*

Responding to stimuli

AQA
Edexcel A   Edexcel B
OCR A ᴬ   OCR A ᴮ
NICCEA
WJEC

KEY POINT   All living organisms can respond to changes in the environment. This is called sensitivity. Plants usually respond more slowly than animals.

Although the speed and type of response may be very different, the order of events is always the same:

stimulus	➤	receptor	➤	co-ordination	➤	effector	➤	response
light, sound, smell, taste or touch		detects the stimulus		usually carried out by the brain or spinal cord		most often a muscle or gland		for example, movement

Fig. 2.18

> Nerves carry messages quicker than hormones.

The **receptors** detect the changes and pass information on to the **central nervous system** (the brain and spinal cord). This coordinates all the information and sends a message to the **effectors** to bring about a response. All these messages are sent by nerves or hormones.

Receptors

AQA
Edexcel A   Edexcel B
OCR A ᴬ   OCR A ᴮ
NICCEA
WJEC

KEY POINT   The job of receptors is to detect the stimulus and send information about it to the central nervous system.

The different receptors in the human body respond to different stimuli:

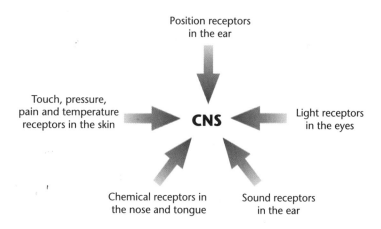

Position receptors in the ear

Touch, pressure, pain and temperature receptors in the skin

CNS

Light receptors in the eyes

Chemical receptors in the nose and tongue

Sound receptors in the ear

Fig. 2.19

> **KEY POINT**
> The receptors are often gathered together into sense organs. They have various other structures that help the receptors to gain the maximum amount of information.

One of these sense organs is the eye:

> **Learn this diagram and how to label it.**

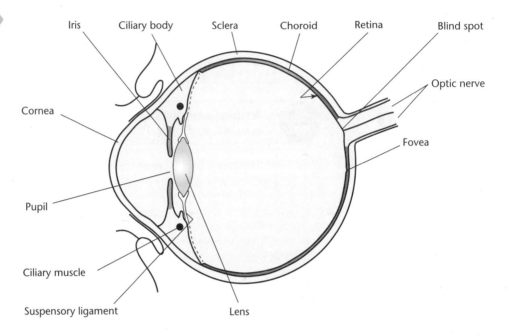

Iris   Ciliary body   Sclera   Choroid   Retina   Blind spot

Optic nerve

Cornea

Fovea

Pupil

Ciliary muscle

Suspensory ligament   Lens

Fig. 2.20

The light enters through the **pupil**. It is focused onto the **retina** by the **cornea** and the **lens**.

The size of the pupil can be changed by the muscles of the iris when the brightness of the light changes. This tries to make sure that the same amount of light enters the eye.

> **Try this by looking in a mirror and turning on and off the light.**

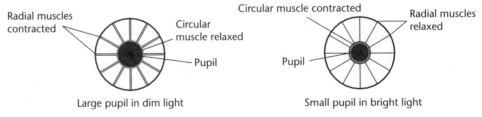

Radial muscles contracted   Circular muscle contracted   Radial muscles relaxed

Circular muscle relaxed

Pupil   Pupil

Large pupil in dim light   Small pupil in bright light

Fig. 2.21

The job of the lens is to change shape so that the image is always focused on the light sensitive retina.

> **KEY POINT**
> To change from looking at a distant object to a near object, the lens has to become more rounded and powerful. This is called accommodation.

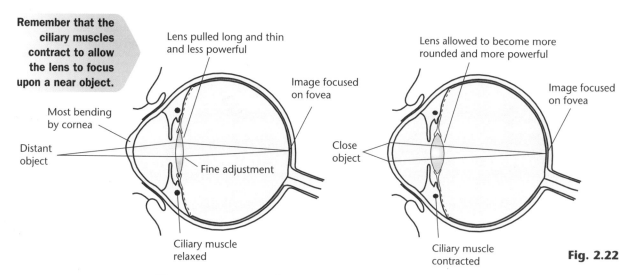

Remember that the ciliary muscles contract to allow the lens to focus upon a near object.

Lens pulled long and thin and less powerful

Image focused on fovea

Most bending by cornea

Distant object

Fine adjustment

Ciliary muscle relaxed

Lens allowed to become more rounded and more powerful

Image focused on fovea

Close object

Ciliary muscle contracted

Fig. 2.22

The receptors are cells in the retina called **rods** and **cones**. They detect light and send messages to the brain along the optic nerve. The rods and cones do slightly different jobs.

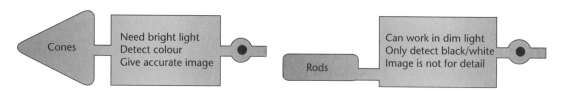

Cones

Need bright light
Detect colour
Give accurate image

Rods

Can work in dim light
Only detect black/white
Image is not for detail

Fig. 2.23

Neurones and responses

AQA
Edexcel A   Edexcel B
OCR A ᴬ   OCR A ᴮ
NICCEA
WJEC

KEY POINT Neurones are specialised cells that carry messages around the body in the form of electrical charges.

There are three main types:

- **sensory neurones** – they carry electrical messages from the sense organs to the CNS

Don't get confused between nerves and neurones. Nerves are collections of thousands of neurones.

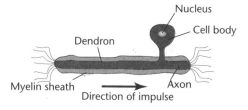

Nucleus

Cell body

Dendron

Myelin sheath

Axon

Direction of impulse

Fig. 2.24

- **motor neurones** – they carry electrical messages from the CNS to the effectors, such as muscles and glands

Make sure that you can put arrows on the neurones to show the direction of the impulse.

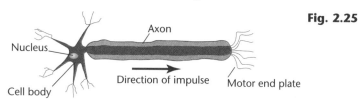

Axon

Nucleus

Cell body

Direction of impulse

Motor end plate

Fig. 2.25

- **relay neurones** – they relay messages between neurones in the CNS.

One neurone does not directly connect with another. The projections at the ends of the neurones end just short of the next neurone. This leaves a small gap.

KEY POINT The junction between two neurones is called a **synapse** and messages are passed across by chemical transmitter molecules.

Many of the drugs mentioned in section 2.8 affect synapses.

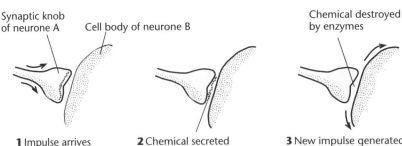

Synaptic knob of neurone A   Cell body of neurone B

Chemical destroyed by enzymes

1 Impulse arrives   **2** Chemical secreted into minute gap (synapse)   **3** New impulse generated by neurone B

Fig. 2.26

Once the information reaches the CNS from a sensory neurone there is a choice:

A. The message can be sent to the higher centres of the brain and the organism might decide to make a response. This is called a **voluntary action**.

B. The message may be passed straight to a motor neurone via a relay neurone. This is much quicker and is called a **reflex action**.

Don't say that reflexes happen unconsciously!

A reflex action	A voluntary action
Very quick, so protects the body	Takes longer
Does not necessarily involve the brain	Always involves the brain
Does not involve conscious thought	Involves conscious thought

KEY POINT A reflex is a rapid response that does not involve conscious thought. It protects the body from damage.

In the withdrawal reflex, pain on the skin is the stimulus. The response is the muscle moving the part of the body away from danger.

1. Stimulus is detected by sensory cell.

2. Impulse passes down sensory neurone.

3. Relay neurone passes impulse to motor neurone.

4. Motor neurone passes impulse to effector.

5. Muscle contracts.

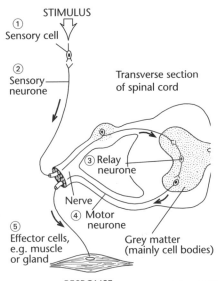

STIMULUS

① Sensory cell

② Sensory neurone

Transverse section of spinal cord

③ Relay neurone

Nerve

④ Motor neurone

Grey matter (mainly cell bodies)

⑤ Effector cells, e.g. muscle or gland

RESPONSE

Fig. 2.27

PROGRESS CHECK

1. What is the job of receptors?
2. What do receptors in the skin detect?
3. How is the shape of the lens in the eye made more rounded?
4. What is the CNS?
5. Name two types of effector organ in the body.

1. Receptors detect stimuli and pass information on to sensory neurones as electrical impulses;
2. Touch, pressure, pain, temperature; 3. The ciliary muscle contracts, loosening the suspensory ligaments; 4. The central nervous system is made up of the brain and spinal cord;
5. Muscles and glands.

2.6 Hormones

LEARNING SUMMARY

After studying this section you should be able to:

● *explain what a hormone is and know where they are produced in the body*
● *understand the role of hormones in controlling reproduction*
● *realise that hormones can be used to control fertility and to improve sporting performances.*

What is a hormone?

AQA
Edexcel A   Edexcel B
OCR A ᴬ   OCR A ᴮ
NICCEA
WJEC

KEY POINT A hormone is a chemical messenger in the body. They are released by glands and pass in the bloodstream to their target organ.

Hormone producing glands are called endocrine glands.

During pregnancy the placenta also makes hormones.

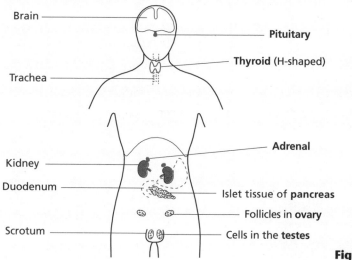

Brain — Pituitary
Thyroid (H-shaped)
Trachea
Adrenal
Kidney
Duodenum — Islet tissue of **pancreas**
Follicles in **ovary**
Scrotum — Cells in the **testes**

Fig. 2.28

The diagram shows the main hormone-producing glands of the body. Between them they make a number of different hormones:

The actions of ADH and insulin are covered in section 2.7.

gland	hormones produced	action
Pituitary	Anti-diuretic hormone (ADH)	Water balance
	Luteinising hormone (LH)	Ovulation and progesterone production
	Follicle stimulating hormone (FSH)	Growth of a follicle
Thyroid	Thyroxine	Controls metabolic rate
Adrenal	Adrenaline	Prepares the body for action
Pancreas	Insulin	Control of blood glucose level
Ovary	Oestrogen	Controls puberty and menstrual cycle in the female
	Progesterone	Maintains pregnancy
Testis	Testosterone	Controls puberty in the male

Hormones and reproduction

Males and females are born with sex organs. They have developed during pregnancy under the influence of the **sex hormones**.

> **KEY POINT**
> At **puberty** the sex hormones are produced in larger amounts and cause further changes, called the **secondary sexual characteristics**.

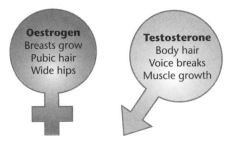

Fig. 2.29

The secondary sexual characteristics also include the production of the sex cells. This process is more complicated in the female because it is not continuous and the hormones must also prepare the uterus to receive a fertilised egg. It involves two hormones from the pituitary gland, FSH and LH.

> Remember: ovulation does not always happen on day 14 of the cycle. Twenty-eight days is only an average length for the cycle. There is a lot of variation between women.

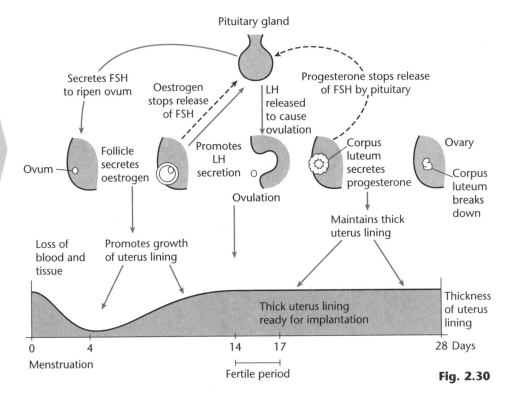

Fig. 2.30

If the egg is fertilised, the embryo sends a message to the ovary stopping the corpus luteum from breaking down. Progesterone production carries on and menstruation does not occur.

Using hormones

AQA

Edexcel A   Edexcel B

OCR A ^A   OCR A ^B

NICCEA

WJEC

KEY POINT | Hormones are often used to try to increase or decrease fertility.

fertility

Many people want to reduce their fertility to avoid unwanted pregnancies.

Oestrogen and progesterone both stop FSH being released by the pituitary gland. They are used in the combined contraceptive pill to stop an egg being developed and released.

> If asked about ways to change fertility remember to include methods to increase and decrease fertility.

fertility

> This may lead to multiple births.

Other people may not ovulate regularly and may use hormones to increase their fertility. FSH or similar hormones can be injected to try to stimulate the ovary to produce eggs.

KEY POINT | Some athletes may take hormones similar to testosterone to try to improve their performance.

> The athletes are tested at random for these drugs throughout the year.

Male sex hormones increase muscle growth and increase aggression. Athletes may use these hormones to improve their strength and make them train harder. These hormones are called **anabolic steroids**.

PROGRESS CHECK

1. What hormone is made in the pancreas?
2. What are the female secondary sexual characteristics?
3. What effect do oestrogen and progesterone have on the production of FSH?
4. Why do pregnant women stop ovulating?
5. What is an anabolic steroid and why do some athletes take them?

1. Insulin;   2. Breast development, widening of the hips, pubic hair and egg production;
3. Inhibit FSH production;   4. The ovary produces high levels of oestrogen and progesterone that will inhibit FSH;   5. An anabolic steroid is a hormone, such as testosterone. Some athletes use them to increase muscle growth and make them train harder.

2.7 Homeostasis

**LEARNING SUMMARY**

After studying this section you will be able to:

- recall that homeostasis is maintaining a constant internal environment
- understand how the kidney controls the urea and water level of the body
- understand how the body controls its own temperature.

Blood glucose

AQA
Edexcel A   Edexcel B
OCR A ᴬ   OCR A ᴮ
NICCEA
WJEC

It is vital that the glucose level in the blood is kept constant. If not controlled, it would **rise** after eating and **fall** when hungry. **Insulin** is the hormone that controls the level of glucose in the blood.

glucose in the blood   insulin   Glycogen in the liver

Fig. 2.31

Remember: glucose is used for respiration to release energy.

KEY POINT   Insulin converts excess glucose into glycogen to be stored in the liver.

Diabetics do not produce enough insulin naturally. They need regular insulin injections in order to control the level of glucose in their blood.

The kidneys – control of waste

AQA
Edexcel A   Edexcel B
OCR A ᴬ   OCR A ᴮ
NICCEA
WJEC

KEY POINT   It is the job of the kidneys to filter urea from the blood.

Urea is a waste material produced from the breakdown of proteins.

The kidney has thousands of fine **tubules**, called **nephrons**.

Blood capillaries carry blood at **high pressure** into these tubules. The small molecules in the blood are squeezed out of the capillaries and collected by the tubules. This is called **ultrafiltration**.

The small molecules that are filtered out include:

water   glucose   salt   urea

If all of this fluid reached the bladder, the whole day would be spent on the loo and drinking water.

The body cannot afford to lose all of this glucose, salt and water. So these molecules are **reabsorbed** back into the blood, leaving just a little water and all of the urea to continue on to the bladder. This liquid is called urine.

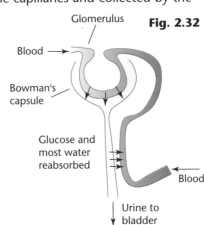

Glomerulus   **Fig. 2.32**

Blood →

Bowman's capsule

Glucose and most water reabsorbed

← Blood

Urine to bladder

The kidneys – control of water

AQA
Edexcel A   Edexcel B
OCR A ᴬ   OCR A ᴮ
NICCEA
WJEC

The kidneys also control the amount of water in our bodies.

Our bodies get water from:

● the food we eat

● the liquids we drink

● water produced by respiration.

Sometimes we do not have enough water.

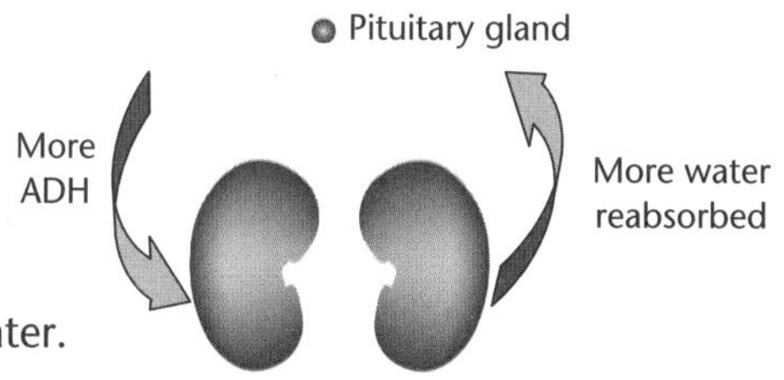

● Pituitary gland

More ADH   More water reabsorbed

Kidneys

Fig. 2.33

Too thirsty?

● the pituitary gland produces more of the hormone called ADH

● this causes the tubules to reabsorb more water

● the urine becomes more concentrated, and the body saves water.

Sometimes we drink too much liquid. Then the opposite happens.

> **Try to think of the body as a container full of liquid. The more water you put in, the more will overflow.**

Drunk too much liquid?

● Less ADH is produced.

● The kidney tubules reabsorb less water.

● More water reaches the bladder.

● Large amounts of dilute urine are produced.

Temperature control

AQA
Edexcel A   Edexcel B
OCR A ᴬ   OCR A ᴮ
NICCEA
WJEC

KEY POINT

● **Our body has a constant temperature of 37°C. This is the temperature at which our enzymes work best.**
● **The average room temperature is about 20°C.**
● **This means we are always warmer than our surroundings.**
● **We are constantly losing heat to our surroundings.**
● **We produce the heat from respiration.**
● **Core body temperature is monitored and controlled by the brain.**
● **Temperature receptors send nerve impulses to the skin.**
● **It is the job of the skin to regulate our body temperature.**

When we feel too hot

> **Common error:**
> **Some students lose marks in exams because they say that blood vessels move away from the surface of the skin. Blood vessels cannot move.**

We feel hot when we need to lose heat faster, as our core body temperature is in danger of rising.

We do this by:

● **sweating** – as water evaporates from our skin, it absorbs heat energy. This cools the skin and the body loses heat.

● **vasodilation** – blood capillaries near the skin surface get wider to allow more blood to flow near the surface. Because the blood is warmer than the air, it cools down and the body loses more heat.

When we feel too cold

When we feel too cold we are in danger of losing heat too quickly and cooling down. This means we need to conserve our heat to maintain a constant 37°C.

We do this by:

> **Remember that blood vessels cannot move.**

- **shivering** – rapid contraction and relaxation of body muscles. This increases the rate of respiration and more energy is released as heat

- **vasoconstriction** – blood capillaries near the skin surface get narrower and this process reduces blood flow to the surface. The blood is diverted to deeper within the body to conserve heat.

PROGRESS CHECK

1. State which hormone converts glucose in the blood into glycogen in the liver.
2. State four small molecules that are filtered from the blood by ultrafiltration.
3. State which of these molecules are not reabsorbed.
4. Explain how the hormone called ADH controls the body's water level.
5. Explain what happens in the skin when we feel too hot.

1. Insulin; 2. Salt, water, glucose, urea; 3. Urea; 4. When thirsty, increased levels of ADH cause more water to be reabsorbed back into the blood. The opposite happens when we drink too much; 5. Sweating causes heat to be lost as the sweat evaporates. Vasodilation of surface capillaries allow more blood to flow to near skin surface, thus allowing the blood to cool down.

2.8 Health

LEARNING SUMMARY

After studying this section you should be able to:

- *understand the main causes of diseases*
- *describe how the body protects itself against disease.*

Causes of disease

AQA
Edexcel A   Edexcel B
OCR A A   OCR A B
NICCEA
WJEC

KEY POINT
A disease occurs when the normal functioning of the body is disturbed. Some diseases can be passed on from one person to another and are called **infectious**.

Genetic diseases are covered in chapter 4.

Type of disease	Description	Examples
Non-infectious:		
Cancer	Uncontrolled cell growth	Lung cancer
Deficiency diseases	Lack of a substance in the diet	Scurvy
Allergies	A reaction to a normally harmless substance	Hayfever
Genetic	Caused by a defective gene (usually inherited)	Sickle cell anaemia
Infectious disease	Caused by a pathogen	Influenza (flu)

Don't use the word 'germ' – examiners do not like it!

There are a number of different organisms that can cause disease.

KEY POINT
Any organism that causes a disease is called a **pathogen**.

Remember: not all bacteria and fungi are pathogens; some are important in the carbon cycle and the nitrogen cycle.

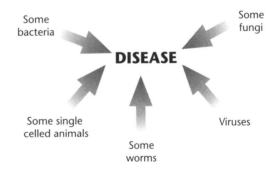

Fig. 2.34

Preventing pathogens from entering the body

AQA
Edexcel A   Edexcel B
OCR A A   OCR A B
NICCEA
WJEC

In order to cause diseases these pathogens need to get into the body. Most are prevented from entering by the **skin**.

KEY POINT
Pathogens often enter through natural openings of the body, such as the mouth, eyes, ears, nose, anus and urethra.

The body has a number of other defences that it uses in order to try to stop pathogens entering it.

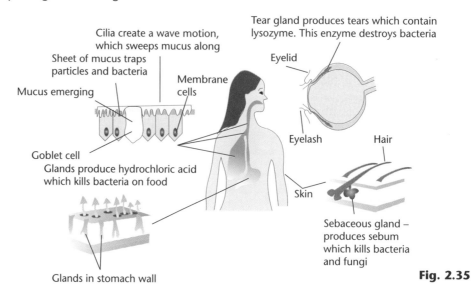

Cilia create a wave motion, which sweeps mucus along

Sheet of mucus traps particles and bacteria

Mucus emerging

Membrane cells

Goblet cell

Glands produce hydrochloric acid which kills bacteria on food

Tear gland produces tears which contain lysozyme. This enzyme destroys bacteria

Eyelid

Eyelash

Hair

Skin

Sebaceous gland – produces sebum which kills bacteria and fungi

Acid in the stomach also provides the best pH for protease to work.

Glands in stomach wall

Fig. 2.35

If the pathogens do enter the body then the body will attack them with two types of white blood cells.

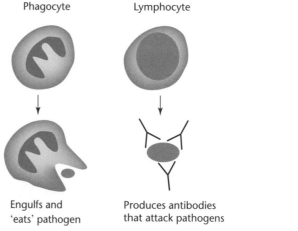

Phagocyte

Lymphocyte

Engulfs and 'eats' pathogen

Produces antibodies that attack pathogens

Fig. 2.36

This is often called a vaccination.

KEY POINT

People may be immunised against a disease by injecting a weakened or dead form of the pathogen into the body. This gives the body advance warning by creating a memory cell and so white blood cells are ready to attack.

Diseases caused by smoking

AQA
Edexcel A   Edexcel B
OCR A ᴬ   OCR A ᴮ
NICCEA
WJEC

Many people cannot give up smoking tobacco because it contains the drug **nicotine**. This alters the action of the nervous system and is highly **addictive**.

KEY POINT

Many drugs are addictive. This means that people keep wanting to use them even though they often have effects called withdrawal symptoms if people stop taking them.

The nicotine may be harmful to the body but most damage is done by the other chemicals in the tobacco smoke:

● Chemicals in the tar may cause cells in the lung to divide uncontrollably. This can cause **lung cancer**.

● The mucus collects in the alveoli and may become infected. This may lead to the walls of the alveoli being damaged. This reduces gaseous exchange and is called **emphysema**.

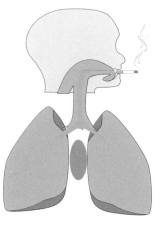

● The heat and chemicals in the smoke destroy the cilia on the cells lining the airways. The goblet cells also produce more mucus than normal. The bronchioles may become infected. This is called **bronchitis**.

● The nicotine can cause an increase in blood pressure increasing the chance of a **heart attack**.

Fig. 2.37

The action of other drugs

AQA
Edexcel A   Edexcel B
OCR A ᴬ   OCR A ᴮ
NICCEA
WJEC

Many drugs have important medical uses but some are misused and can have dangerous side effects on the body. Drugs often act on the nervous system by affecting synapses. They may have a number of actions:

The action of synapses is covered in section 2.5.

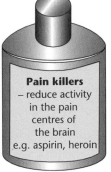

Sedatives – slow down the action of the nervous system e.g. alcohol, barbiturates

Pain killers – reduce activity in the pain centres of the brain e.g. aspirin, heroin

Stimulants – increase the activity of the nervous system e.g. caffeine, amphetamines

Fig. 2.38

PROGRESS CHECK

1. What is a pathogen?
2. How does mucus help to stop pathogens from entering the body?
3. What are lymphocytes and what do they do?
4. Write down two diseases of the lungs that are more likely to occur in people who smoke tobacco.

1. An organism that causes disease;   2. The mucus is sticky and this traps microbes;   They are then wafted up by cilia to the mouth and swallowed;   3. Lymphocytes are a type of white blood cell. They make antibodies;   4. Two from: emphysema, lung cancer, bronchitis.

Sample GCSE question

1.(a) David was running the 5000m in his school's sports day. He noticed that after a few seconds his heart was beating faster. Explain why his heart beat increased as he started to run. **[3]**

> To pump more blood to the muscles ✓ so that they could get more oxygen ✓ and glucose ✓.

Respiration starts off as aerobic but changes to anaerobic as the supply of oxygen is exceeded by the demand.

(b) He also noticed that as he ran his breathing rate increased. Explain why David breathed faster. **[3]**

> To absorb more oxygen ✓ for respiration ✓ and to get rid of excess carbon dioxide ✓.

If you cannot remember the numbers of atoms, start to balance the equation with the carbon atoms first, then hydrogen and finally oxygen.

(c) Complete the following equation to show how David was supplying his muscles with energy. **[2]**

$$C_6H_{12}O_6 + 6O_2 \rightarrow 6H_2O + 6CO_2 \checkmark \checkmark$$

(d) After a couple of minutes David's muscles started to run short of oxygen. They broke down the glucose without oxygen. State what this type of respiration is called. **[1]**

> Anaerobic ✓.

Sometimes you are asked to write the word equation rather than the chemical one. If you choose to write down the chemical equation instead, be warned. If you make a mistake you will lose marks on what should have been an easy question.

(e) Complete the word equation to show what was happening in David's muscles. **[1]**

> Glucose → lactic acid ✓.

You can include the word 'energy' in this equation if you wish.

(f) When the race was over David noticed that he was still out of breath even though he had stopped running. Explain why David was still panting. **[3]**

> Lactic acid had built up in his muscles ✓. It is toxic and must be broken down ✓. Oxygen is required for this and David continues to pant until he has absorbed enough oxygen to break down all the lactic acid ✓.

Remember that this is called the 'oxygen debt'.

(g) The glucose that David used in his race came from his breakfast. Explain how the carbohydrate that David ate got into his bloodstream as glucose. **[3]**

> Carbohydrate digested by amylase, in his gut ✓, which is alkaline ✓. The glucose molecules are then absorbed through the gut into the bloodstream ✓.

(h) Once the glucose entered David's blood, it was transported to the muscles in his legs. Describe the journey taken by the glucose as it goes from the gut, through the heart and eventually to the leg muscles. **[2]**

> Gut → heart → lungs ✓ → heart ✓ → leg muscles.

The two marks are for realising that the blood goes to the lungs and back to the heart.

Exam practice questions

1. Joy has 5 litres of blood. The blood is filtered by her kidneys 250 times every day.

(a) State how many litres of blood are filtered by Joy's kidneys each day. **[2]**

Joy makes a model of her kidney. She puts a cellophane bag of artificial blood into a beaker of distilled warm water.

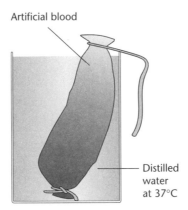

Artificial blood

Distilled water at 37°C

(b) She tests the artificial blood and the distilled water, to see if any of the substances have leaked out of the bag.

	Joy's results		
substance	**found in blood**	**found in water**	**not found in water**
glucose	✓	✓	
salt	✓	✓	
protein	✓		✓
urea	✓	✓	

(i) Explain why Joy kept the distilled water at 37°C. **[1]**

(ii) Name one substance which did not pass out of the bag. **[1]**

(iii) Explain why this substance did not pass out of the bag. **[1]**

(c) Real blood contains red cells and white cells.
Explain the job carried out by each of these types of cell. **[2]**

(d) Joy knows that when kidneys stop working, a person has to use a kidney machine, or have a transplant. Joy's mother carries a kidney donor card but her father does not agree with them.

(i) Suggest two problems faced by a person who has to use a kidney machine. **[2]**

(ii) Explain whether you think that Joy's father should be made to carry a donor card by law. **[3]**

Exam practice questions

2.

(a) The diagram shows the outline of Jack's body.

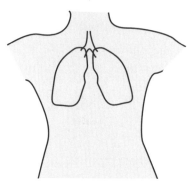

    (i) State the name of the organ shown in the diagram. **[1]**

    (ii) Mark the position of Jack's heart by drawing a ♥ on the diagram. **[1]**

    (iii) Mark the position of Jack's stomach by drawing a ◯ on the diagram. **[1]**

(b) Jack starts to run a race. Aerobic respiration takes place in his cells.

Complete the word equation for aerobic respiration.

glucose + _____ → _____ + water + energy **[2]**

(c) When the race is nearly over, Jack's muscles are tired and painful.

    (i) State what type of respiration is taking place now. **[1]**

    (ii) Complete the word equation for this type of respiration.

glucose → _____ + energy **[1]**

(d) Describe how Jack used his diaphragm and intercostal muscles, to draw air into his lungs. **[3]**

The diagram shows how oxygen diffuses from Jack's lungs, into his blood stream. The artist forgot to draw the oxygen molecules in Jack's blood.

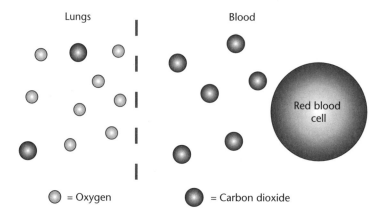

(e) Draw the oxygen molecules in Jack's blood, left out by the artist. **[2]**

Green plants as organisms

The following topics are covered in this section:

- **Green plants as organisms**
- **Plant hormones**
- **Transport in plants**
- **Support**
- **Sugar transport**

What you should know already

Finish the passages using words from the list. These words should also be used to complete the equation. You may use the words more than once.

atmosphere	carbon dioxide	light	mineral	nitrogen	oxygen
photosynthesis	root	same	sugar	water	wilts

Plants grow by absorbing 1._____ energy from the Sun. This process is called 2._____. They use the energy to combine water and 3._____ to make food and a gas called 4._____.They release this gas into the 5._____ and we are then able to breathe it in. Plants also need the 6._____, so that they can convert the food that they make into protein.

$$water + 7.\underline{\hspace{3cm}} \rightarrow sugar + 8.\underline{\hspace{2cm}}$$

Plants also carry out respiration all of the time. This means that at night they remove the gas 9._____ from the air and reduce the levels of stored 10._____ in the plant. This is why some people wrongly think that flowers should not be left in a sick person's bedroom, at night. If you write the word equation for respiration backwards, it is the 11._____ as the equation for photosynthesis.

Root hairs

Plants roots are covered in very tiny root hairs. They are so small and thin that they can absorb water and 12._____ salts straight from the soil. The water enters the root hair and is then passed from cell to cell until it reaches the centre of the 13._____.

Root hairs are so delicate, that when we dig up a plant to move it somewhere else, the hairs are broken and damaged and the plant cannot absorb water through them. This is why the plant 14._____, until it has grown some more root hairs and can once more absorb 15._____ from the soil.

ANSWERS

1. light; 2. photosynthesis; 3. carbon dioxide; 4. oxygen; 5. atmosphere; 6. nitrogen; 7. carbon dioxide; 8. oxygen; 9. oxygen; 10. sugar; 11. same; 12. mineral; 13. root; 14. wilts; 15. water

3.1 Green plants as organisms

After studying this section you should be able to:

● *understand photosynthesis*
● *understand about limiting factors*
● *understand the role of mineral salts.*

Nutrition

Photosynthesis

Photosynthesis is the process where plants make the food glucose, from carbon dioxide and water. It uses the energy in sunlight and the green pigment chlorophyll, found in chloroplasts.

The equation

> Remember: the equation for respiration is the equation for photosynthesis backwards.

Carbon dioxide + water $\xrightarrow[\text{Chlorophyll}]{\text{Light}}$ glucose + oxygen

Fig. 3.1

The balanced equation for photosynthesis is:

$$6CO_2 + 6H_2O \rightarrow C_6H_{12}O_6 + 6O_2$$

The leaf

The process of photosynthesis takes place mainly in the leaf.

> The leaf is ideally adapted to its job. It is thin, light and has a large surface area.

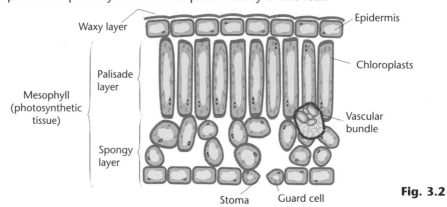

Waxy layer — Epidermis
Palisade layer — Chloroplasts
Mesophyll (photosynthetic tissue)
Spongy layer — Vascular bundle
Stoma   Guard cell

Fig. 3.2

Chloroplasts are mainly found near the upper surface of the leaf. They absorb the energy from sunlight in order to power the reaction.

Stomata are found on the under surface of the leaf. They open during the day to absorb carbon dioxide and release oxygen. They close at night in order to stop the loss of water.

The **vascular bundles** contain **xylem** vessels, which transport water and **phloem** vessels, which transport glucose.

Photosynthesis versus respiration

You must remember that respiration continues all the time.

KEY POINT Photosynthesis only occurs during the hours of daylight. Plants respire all of the time.

A common error is that many students think that plants only respire at night.

However, during the day, photosynthesis proceeds much faster than respiration, so it is easy to see why some students make this mistake.

There are two times during the day when photosynthesis and respiration are equal. At these times, the carbon dioxide being used by photosynthesis is equal to the carbon dioxide being produced by respiration.

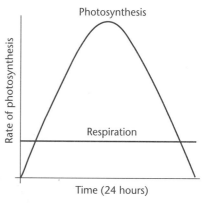

Fig. 3.3

Limiting factors

AQA
Edexcel A   Edexcel B
OCR A ^A   OCR A ^B
NICCEA
WJEC

Photosynthesis is a chemical reaction. The rate, or speed of this reaction is limited by the following factors:

- shortage of light
- shortage of carbon dioxide
- low temperature.

KEY POINT These three factors are called limiting factors because they limit the rate of photosynthesis.

Exam questions on limiting factors usually involve graphs like these.

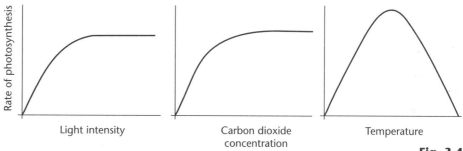

Fig. 3.4

The graphs show that as light and carbon dioxide increase, so does the rate of photosynthesis, until the rate levels out at a new optimum level. The rate is then stable until the new limiting factor is removed. **Temperature** is different – any increase above the optimum level causes the rate to slow and stop. This is because high temperature **denatures** the enzymes.

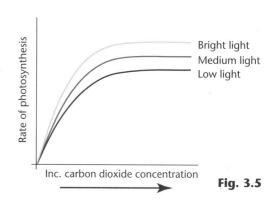

Fig. 3.5

What next?

Once glucose has been made in the chloroplasts, many different things can happen to it.

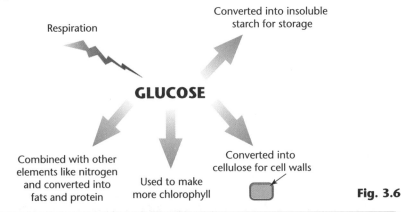

Respiration

Converted into insoluble starch for storage

GLUCOSE

Combined with other elements like nitrogen and converted into fats and protein

Used to make more chlorophyll

Converted into cellulose for cell walls

Fig. 3.6

Mineral salts

 AQA
 Edexcel A Edexcel B
OCR A ^A OCR A ^B
 NICCEA
WJEC

For healthy growth plants also need mineral salts:

- **nitrates** to make proteins
- **phosphates** for DNA, cell membranes and chemical reactions
- **potassium** to help enzymes to work
- **magnesium** to make chlorophyll.

Fig. 3.7

A lack of these minerals has serious effects on the plants.

no nitrates
stunted growth
yellow older leaves

no phosphates
purple younger
leaves

no potassium
yellow leaves with
dead spots

no magnesium
stunted growth
pale yellow leaves

 PROGRESS CHECK

1. Write out a balanced equation for photosynthesis.
2. How many molecules of glucose are made from six molecules of water and six molecules of carbon dioxide?
3. Explain how a leaf is adapted to its function.
4. On a normal sunny day, how many times does the rate of photosynthesis equal the rate of respiration?
5. List three limiting factors for photosynthesis and explain which is the odd one out.
6. State five things that can happen to glucose after it is made during photosynthesis.
7. Name four mineral salts needed by plants and for each one explain how the plant uses it and what happens when it is missing.

1. $6CO_2 + 6H_2O \rightarrow C_6H_{12}O_6 + 6O_2$; 2. One; 3. Thin, light, large surface area, has vascular bundles for transport and stomata to absorb carbon dioxide and release oxygen; 4. Two; 5. Light, carbon dioxide and temperature. Temperature is the odd one out because if too high enzymes are denatured; 6. Converted to starch, protein, cellulose. Used to make more chlorophyll or for respiration; 7. Nitrates for proteins. Stunted growth and yellow older leaves. Phosphates for chemical reactions. Stunted growth and purple younger leaves. Potassium to help enzymes. Yellow leaves with dead spots. Magnesium to make chlorophyll. Stunted yellow leaves.

3.2 *Plant hormones*

LEARNING SUMMARY

After studying this section you should be able to:

● **understand how plants control their growth**
● **understand the role played by hormones in the process**
● **state some commercial uses of plant growth hormones.**

Control of plant growth

AQA
Edexcel A | Edexcel B
OCR A ᴬ | OCR A ᴮ
NICCEA
WJEC

Just like humans, plants also respond to a stimulus, although much more slowly. They do it by growing towards or away from the stimulus:

● **phototropism** – shoots respond to light by growing towards it
● **geotropism** – shoots grow away from gravity, roots grow towards it
● **hydrotropism** – roots grow towards water.

KEY POINT Auxins are plant growth hormones.

How auxins work

Fig. 3.8

Auxins make cells grow longer.

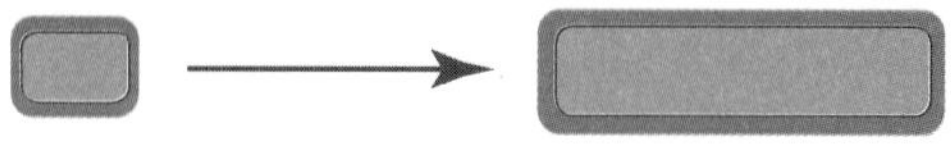

Shoots grow towards the light because auxins collect on the dark side of the shoot. This causes the cells on the dark side to lengthen.

Auxin collects on dark side

Auxin collects on dark side and the shoot lengthens and bends towards the light

Fig. 3.9

> *These techniques are often used by commercial growers to increase their productivity.*

Other uses for plant growth hormones:

● hormone rooting powder promotes the growth of roots in shoot cuttings
● unpollinated flowers can be treated to produce seedless fruits
● ripening of fruits can be slowed down in order to keep them fresh during transport to consumers
● some weedkillers contain a synthetic hormone which causes broad leaf plants to 'outgrow' themselves and die.

PROGRESS CHECK

1. Describe three different plant growth responses controlled by hormones.
2. Explain how the hormone auxin, causes a plant shoot to grow towards the light.
3. State four ways that a commercial grower might use plant growth hormones to improve his crops.

1. Phototropism – shoots grow towards light   Geotropism – roots grow towards gravity and shoots away from gravity   Hydrotropism – roots grow towards water;   2. Auxin accumulates on dark side of shoot. Causes cells on dark side to grow longer. This causes shoot to bend towards light;   3. Root cuttings, seedless fruits, delay ripening, weedkiller.

3.3 Transport in plants

After studying this section you should be able to:

- *understand how trees get water to their highest branches*
- *know why plants transport water up to their leaves*
- *know how plants support themselves without a skeleton*
- *know how plants transport food.*

Transpiration

> Osmosis is explained in section 1.2

Water enters the root

Water enters the plant through the root hairs, by osmosis. It then passes from cell to cell, by osmosis until it reaches the centre of the root.

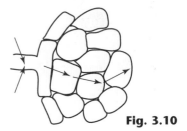

Fig. 3.10

Water goes up the stem

The water enters xylem vessels in the root, and then travels up the stem. Xylem vessels are dead tubular cells connected together. They are hollow and the ends of the cells have been removed.

Xylem

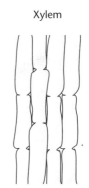

Fig. 3.11

Water evaporates from the leaves

Fig. 3.12

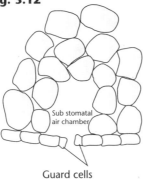

Sub stomatal air chamber

Guard cells

Water evaporates from the cells and the vapour collects in the sub-stomatal air chamber. It then passes through the stomata by diffusion.

Factors affecting the rate of transpiration:

- **temperature** – warm weather increases the kinetic energy of the water molecules so they move out of the leaf faster

- **humidity** – damp air reduces the concentration gradient so the water molecules leave the leaf more slowly

- **wind** – the wind blows away the water molecules so that a large diffusion gradient is maintained.

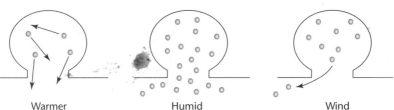

Warmer   Humid   Wind   **Fig. 3.13**

How does water manage to get to the top of tall trees?

Is it pulled or is it pushed?

When water enters the root by osmosis, the maximum height that can be reached is about 10 metres. Some trees can be 100 metres high.

> **KEY POINT** **Water is pulled up the tree by transpirational pull.**

Because:

- water molecules stick to each other by cohesion
- water molecules stick to the walls of the xylem by adhesion
- as water molecules evaporate from the leaves, a thread of water is pulled up from the roots
- cohesion and adhesion stop this thread from stretching or snapping.

The water problem – and how to solve it

When plants are short of water, they do not want to waste it through transpiration.

- The purpose of the stomata is not to lose water, but to let in carbon dioxide. Photosynthesis only occurs during the day, so the stomata close at night to reduce water loss.

Fig. 3.14

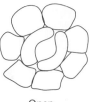 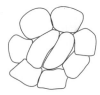

Open   Closed

> **They have various mechanisms for reducing the amount of water they lose.**

- The stomata are placed on the underside of the leaf. This reduces water loss because they are away from direct sunlight and protected from the wind.
- The top surface of the leaf, facing the Sun, is often covered with a protective waxy layer.

Uptake of mineral salts

Fig. 3.15

Minerals are taken into plants in one of two ways:

- **diffusion** as they are swept along with the flow of water down a concentration gradient.
- **active transport** – some minerals are actively transported into the cell. This requires energy.

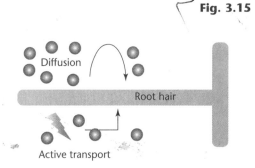

Diffusion

Root hair

Active transport

3.4 Support

After studying this section you should be able to:

- understand that plants do not have a skeleton
- understand that trees use dead wooden cells for support
- understand that smaller plants do not have any wood.

How water supports a plant

AQA
Edexcel A   Edexcel B
OCR A ^{A} OCR A ^{B}
NICCEA
WJEC

When a bicycle tyre is inflated with air, it gets hard. Plants use this principle to make their cells hard.

This is why most fruit and vegetables are crunchy.

- Plant cells (unlike animal cells) are enclosed in a cellulose box, called a cell wall.

- When water enters the plant cells by osmosis, the cell membrane is pushed hard against the cellulose wall.

- The cellulose wall cannot expand, so the cell gets harder. This is called turgor.

If the cell loses too much water, the membrane pulls away from the cell wall and the cell is plasmolysed.

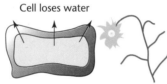

Cell loses water

Cell collapses and plant wilts

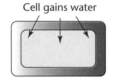

Cell gains water

Cell turgid and hard

Fig. 3.16

3.5 Sugar transport

After studying this section you should be able to:

- understand that glucose sugar is made in the leaf by photosynthesis
- understand that most plants store sugar by converting it into starch
- understand that the starch is usually stored in places like roots, which are a long way from the leaves
- understand that plants need to transport sugar from the leaves to the growing regions or storage areas.

Phloem

AQA
Edexcel A   Edexcel B
OCR A ^{A} OCR A ^{B}
NICCEA
WJEC

KEY POINT   Plants transport dissolved sugar through vessels called phloem.

Phloem and xylem are grouped together in vascular bundles.

● Unlike xylem, phloem cells are living and full of cytoplasm.

● Each cell is joined to the next by holes that connect the cytoplasm together.

● The cytoplasm forms a continuous system of living material to transport the sugar.

Phloem

Fig. 3.17

> **KEY POINT**
>
> **Food is transported as sugar because sugar is soluble and stored as starch because starch is insoluble. This ensures that the stored food stays where it is put.**

PROGRESS CHECK

1. Does water get to the top of tall trees because it is pushed or pulled?
2. State the two forces that ensure that the thread of water moving up the tree does not snap?
3. State three environmental factors that would increase the rate of transpiration.
4. State three features of a xylem vessel.
5. Explain the difference between how water and mineral salts, enter a root hair.
6. Explain how a leaf manages to reduce water loss by transpiration.
7. Explain the role of water in supporting plants.
8. State three differences between xylem and phloem.
9. Explain why food in plants is usually stored as insoluble starch.

1. Pulled; 2. Cohesion and adhesion; 3. Warmer, drier, windier; 4. Dead, empty, cells have no ends; 5. Water enters by osmosis, mineral salts enter by diffusion and active transport; 6. Stomata can close, are on underside of leaf, waxy layer on upper surface of leaf; 7. Water enters cell by osmosis, cell pushes against cell wall, cell gets hard; 8. Xylem dead, no ends to cell, empty – phloem alive, has holes in ends of cell, contains cytoplasm; 9. Stored food cannot leave the cell if it is insoluble.

Sample GCSE question

1. Plants absorb energy from sunlight to make food, by photosynthesis.

(a) State the equation for photosynthesis. [3]

$$6H_2O + 6CO_2 \rightarrow C_6H_{12}O_6 + 6O_2 \checkmark\checkmark\checkmark$$

This is the equation for respiration backwards. One mark for correct balancing, one mark for correct reactants and one mark for correct products.

(b) Explain why chloroplasts that absorb the light energy are found mainly near the upper surface of the leaf, while stomata, that absorb carbon dioxide, are found near the lower surface. [3]

The chloroplasts are situated near the surface of the leaf because that is where the light energy is at its greatest ✓. Stomata however are situated on the under side of the leaf because they also lose water ✓. It is cooler and there is less air movement underneath the leaf so less water is lost by transpiration ✓.

One mark for light energy being greater near upper surface. One mark for stomata losing water and another mark for the explanation.

(c) Plants photosynthesise during daylight hours, but respire all of the time. Explain why plants manage to produce a surplus of glucose when they spend more time respiring than they do photosynthesizing. [3]

During daylight hours the rate of photosynthesis is much greater than the rate of respiration ✓. Although at night respiration is greater than photosynthesis, respiration occurs at a much lower rate ✓. So over a 24 hour period, photosynthesis exceeds respiration ✓.

You could also refer to the compensation point, which occurs twice a day, when photosynthesis and respiration are equal.

(d) Market gardeners know that they can increase crop production in greenhouses by raising the level of carbon dioxide. Look at the following graph.

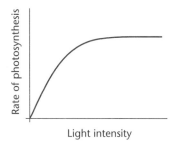

(i) State what other limiting factor it shows. [1]

Light ✓.

(ii) Suggest what the gardener could do to increase production even more. [2]

Raise light levels even higher ✓. Increase temperature ✓.

Questions that start with 'suggest' are asking for your ideas, not your knowledge.

Exam practice questions

1. A group of scientists lived in a self-contained dome called a biosphere.

    They managed to grow enough food to be self-sufficient. The crops that they grew also provided them with enough oxygen for them to breathe.

    (a) Name the process by which plants make food. [1]

    (b) Glucose is made from carbon, hydrogen and oxygen.

         Explain where plants get the carbon from. [2]

    (c) The diagram shows a section through a leaf taken from the biosphere.

    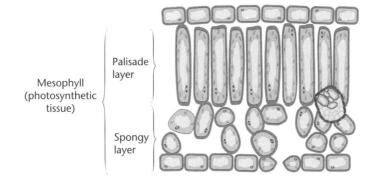

         (i)   Draw an arrow on the diagram to show where the leaf puts oxygen into the atmosphere. [1]

         (ii) Label those structures that are responsible for trapping the energy in sunlight. [1]

         (iii) Explain why these structures are mostly found near the upper surface. [1]

    (d) The scientists knew that the plants would not put oxygen into the atmosphere during the night.

         Explain why the scientists were concerned that the plants might actually remove oxygen from the air, during the night. [2]

    (e) The experiment was stopped because the level of oxygen in the biosphere began to drop. The reason was that not enough light was entering the biosphere.

         Explain why a lack of light would reduce the oxygen level. [2]

    (f) One of the scientists said that carbon dioxide was a limiting factor and increasing its level in the biosphere would have increased the rate of photosynthesis.

         Explain why he thought this. [3]

2. Mark grew some tomato plants in the conservatory. He noticed that they were all leaning towards the light.

    (a) State the name of the process where plants grow towards the light. [1]

         Mark knew that plants were able to do this because of a hormone called auxin.
         The diagram (on page 58) shows auxin in the stem of one of his tomato plants.

Exam practice questions

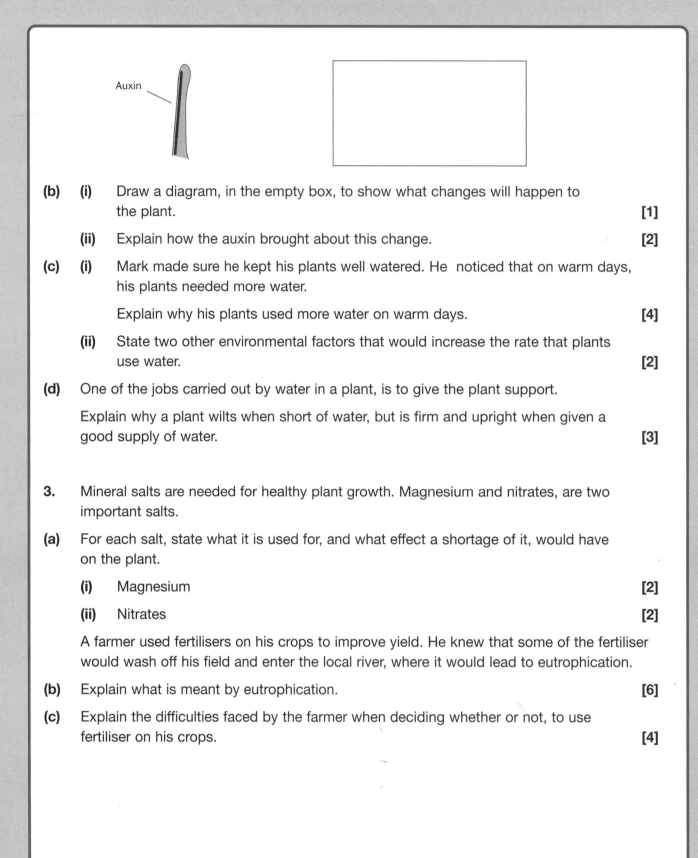

Auxin

(b) **(i)** Draw a diagram, in the empty box, to show what changes will happen to the plant. **[1]**

(ii) Explain how the auxin brought about this change. **[2]**

(c) **(i)** Mark made sure he kept his plants well watered. He noticed that on warm days, his plants needed more water.

Explain why his plants used more water on warm days. **[4]**

(ii) State two other environmental factors that would increase the rate that plants use water. **[2]**

(d) One of the jobs carried out by water in a plant, is to give the plant support.

Explain why a plant wilts when short of water, but is firm and upright when given a good supply of water. **[3]**

3. Mineral salts are needed for healthy plant growth. Magnesium and nitrates, are two important salts.

(a) For each salt, state what it is used for, and what effect a shortage of it, would have on the plant.

(i) Magnesium **[2]**

(ii) Nitrates **[2]**

A farmer used fertilisers on his crops to improve yield. He knew that some of the fertiliser would wash off his field and enter the local river, where it would lead to eutrophication.

(b) Explain what is meant by eutrophication. **[6]**

(c) Explain the difficulties faced by the farmer when deciding whether or not, to use fertiliser on his crops. **[4]**

Variation, inheritance and evolution

The following topics are covered in this section:

- ● **Variation**
- ● **Inheritance**
- ● **Evolution**

What you should know already

Finish the passage using words from the list. The words should also be used to label the classification diagram. You can use the words more than once.

amphibians	animal	bird	fish	flowering	invertebrates
mammals	non-flowering	plant	reptile	vertebrates	

Humans belong to a group of warm blooded animals called 1._____. Humans, birds and reptiles, all have backbones. Animals without backbones are called 2._____. An example of a cold blooded vertebrate is a 3._____. A fir tree is a 4._____ plant, whereas a rose is a 5._____ plant.

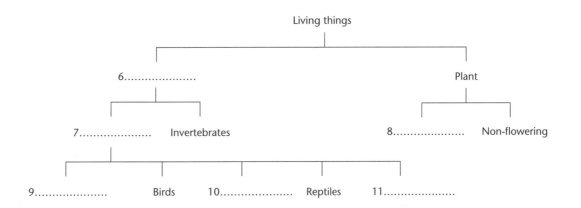

```
                           Living things
          ┌──────────────────┴──────────────────┐
   6...................                        Plant
      ┌───────┴───────┐                  ┌───────┴───────┐
7................  Invertebrates    8................  Non-flowering
   ┌──────┬──────┬──────┬──────┐
9...........  Birds  10...........  Reptiles  11...........
```

Organisms look different from one another because of variation. Some variations are inherited, and some are caused by the environment.

Choose from the following list to give two examples of inherited variation in humans, and two examples of environmental variation.

blood group   **eye colour**   **intelligence**   **sun tan**   **tattoo**

Inherited 12._____ 13._____.

Environmental 14._____ 15._____.

State which of the examples given is a combination of both genetic and environmental factors.

16._____.

4.1 Variation

After studying this section you should be able to:

● **explain the causes of variation**
● **understand the role of sexual and asexual reproduction in variation**
● **understand what a mutation is**
● **explain the causes of mutations.**

Causes of variation

AQA
Edexcel A   Edexcel B
OCR A ᴬ   OCR A ᴮ
NICCEA
WJEC

Children born from the same parents all look slightly different. We call these differences 'variation':

● **inherited or genetic** – some variation is inherited from our parents

● **environmental** – some variation is a result of our environment.

> Remember:
> variation can arise
> in two ways.

Examples of different kinds of variation	
Inherited	**Environmental**
eye colour	sun tan
blood group	scar tissue
finger prints	tattoos
hair colour	hair length
height and weight	
These can be a combination of both inherited and environmental causes.	

A good way to think of it, is that the genes provide a height and weight range into which we will fit, and how much we eat determines where in that range we will be.

Scientists have argued for many years whether 'nature' or 'nurture' (inheritance or environment), is responsible for intelligence.

Nature **INTELLIGENCE** Nature

Fig. 4.1

A scientist called Francis Galton thought that intelligence was inherited and that the environment had nothing to do with it. The argument can be resolved by studying identical twins that have been separated at birth.

Genetically, both twins are the same. Therefore any differences between them must be due to the environment. Tests on identical twins tell us that intelligence is a mixture of both our genes and the environment.

How sexual reproduction leads to variation

AQA
Edexcel A   Edexcel B
OCR A ᴬ   OCR A ᴮ
NICCEA
WJEC

Sexual reproduction involves the joining together of male and female gametes.

The gametes contain chromosomes on which are found genes. Genes are the instructions that make an organism.

Mum and dad like all other humans have 46 (23 pairs) chromosomes in most of the cells of their bodies. This is called the **diploid** number.

> This type of cell division is called meiosis.

Males produce sperm that contain 23 chromosomes. One from each pair.

Females produce ova that contain 23 chromosomes. One from each pair. This is called the **haploid** number.

When a sperm fertilises an ovum, the number returns to 46 (23 pairs).

If this did not happen, the number of chromosomes would double with each generation.

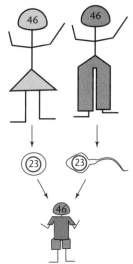

Fig. 4.2

Because the baby can receive any one of the 23 pairs from mum and any one of the 23 pairs from dad, the number of possible gene combinations is enormous. This new mixture of genetic information produces a great deal of variation in the offspring.

How asexual reproduction leads to clones

> This type of cell division is called mitosis.

Asexual reproduction is when cells divide to make identical copies of themselves. The number of chromosomes stays the same. Plants and animals do this when they grow.

Some plants grow so much that they produce smaller plants. This is called asexual reproduction, as there is no sex involved.

Fig. 4.3

Only one individual is needed for asexual reproduction.

The 'baby' plants are genetically identical to their parents. They are called **clones**. When you go into a shop and buy some strawberries, all the strawberries are genetically identical. You are eating strawberry clones.

Mutation – a source of variation

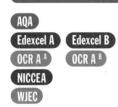

A mutation is a random change to the structure of DNA. DNA has a language just like English, but whereas English has 26 letters, DNA has just 4 chemical 'letters'. These four 'letters' are called bases.

A mutation is when one of these chemical 'letters' is changed. When this happens, it is most unlikely to benefit the organism.

Think what would happen if you made random changes to a few of the letters on this page. It is most likely to produce gibberish and very unlikely to make any sense at all.

- If a mutation occurs in a gamete, the offspring may develop abnormally and could pass the mutation on to their own offspring.

- If a mutation occurs in a body cell, it could start to multiply out of control – this is cancer.

> On very rare occasions; however, a single random mutation can make a major change to what the gene is saying.

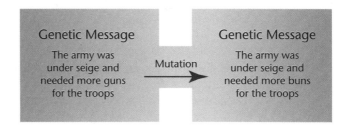

Genetic Message		Genetic Message
The army was under seige and needed more guns for the troops	Mutation →	The army was under seige and needed more buns for the troops

Fig. 4.4

On the rare occasions when a beneficial mutation occurs, natural selection ensures that it will increase in the population.

Causes of mutations:

> Anything that changes or damages the genes on the DNA can cause a mutation.

- radiation
- UV in sun light
- X-Rays
- chemical mutagens – as found in cigarettes.

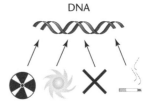

DNA

Fig. 4.5

PROGRESS CHECK

1. Which of the following examples of variation is inherited?
   sun tan   tattoo   eye colour   scar tissue
2. Who was the scientist that thought intelligence was inherited?
3. Which type of reproduction produces variation in the offspring?
4. Name one other source of variation.
5. List four things that can cause mutations.
6. Why do mum and dad have 46 chromosomes, but produce sperm and ova with only 23?
7. What type of cell division leads to the production of clones?

1. Eye colour;   2. Francis Galton;   3. Sexual reproduction;   4. Mutation;   5. Radiation, UV, X-Rays, Chemical mutagens – as found in cigarettes;   6. So that after fertilisation, the number in the baby returns to 46;   7. Mitosis.

4.2 Inheritance

After studying this section you should be able to:

LEARNING SUMMARY

● **explain how sex is determined**
● **understand monohybrid inheritance**
● **explain how some diseases are inherited**
● **understand the gene**
● **understand cloning, selective breeding and genetic engineering.**

Sex determination

AQA
Edexcel A   Edexcel B
OCR A A   OCR A B
NICCEA
WJEC

Humans have 23 pairs of **chromosomes**. The chromosomes of one of these pairs are called the sex chromosomes because they carry the genes that determine the sex of the person.

● Females have two X chromosomes and are XX.

● Males have an X and a Y chromosome and are XY.

KEY POINT There are two kinds of sex chromosome. One is called X and one is called Y.

This means that females produce ova that contain single X chromosomes and males produce sperm, half of which contain a Y chromosome and half of which contain an X chromosome.

Boys inherit an X chromosome from their mother and a Y chromosome from their father.

Fig. 4.6

Girls inherit an X chromosome from their mother and an X chromosome from their father.

	X	Y
X	XX	XY
X	XX	XY

Monohybrid inheritance

AQA
Edexcel A   Edexcel B
OCR A A   OCR A B
NICCEA
WJEC

Gregor Mendel was an Augustinian monk who did experiments with pea plants. He formulated two laws:

● **Law of segregation** – the alleles of a gene separate into different gametes

● **Law of independent assortment** – any male gamete can fertilise any female gamete.

Mendel could not back up his ideas with science as the technology of microscopes had not yet been discovered. It was many years before the science of modern genetics was born.

We now know that we inherit 23 chromosomes from mum and 23 chromosomes from dad.

> **KEY POINT** This means that we each get two sets of instructions.

Each set of 23 is a complete set of instructions for making us. Therefore, each gene or instruction has two alleles: one comes from mum and the other one comes from dad.

A good example to explain this is tongue rolling. The two alleles for tongue rolling are:

● YES – you can roll your tongue

● NO – you cannot roll your tongue.

This means that the possible combinations we can inherit are:

Allele from mum	Allele from dad	What the baby gets
YES	YES	YES YES
YES	NO	YES NO
NO	YES	YES NO
NO	NO	NO NO

> This means that only people with NO NO will *not* be able to roll their tongue.

If the alleles agree with each other there is no problem, but sometimes the alleles disagree about tongue rolling. When this happens, tongue rolling is always **dominant** and non-tongue rolling is always **recessive**.

Instead of using YES and NO we use a capital **T** for tongue rolling and a lower case **t** for non-tongue rolling.

> **KEY POINT** Words you need to know:
> homozygous – both alleles agree
> heterozygous – both alleles disagree
> genotype – which type of alleles make up the gene
> phenotype – how the gene expresses itself.

Some examples

Mum and dad are both **homozygous**. Dad's **phenotype** is a tongue roller. Mum's phenotype is a non-tongue roller.

> All the children are hetrozygous tongue rollers.

	mum	
	t	t
dad T	Tt	Tt
dad T	Tt	Tt

← All are tongue rollers

In this example, both mum and dad are **heterozygous** and have the phenotype of a tongue roller.

> Three out of four can roll their tongues. One out of four cannot roll their tongues. This is a 3:1 ratio.

	mum	
	T	t
dad T	Tt	Tt
dad t	Tt	tt

← Cannot roll tongue

Inherited diseases

AQA
Edexcel A   Edexcel B
OCR A ᴬ   OCR A ᴮ
NICCEA
WJEC

Diseases We Catch
Measles
Flu
Chicken pox

Diseases caused by faulty Genes & Chromosomes
Cystic fibrosis
Huntington's chorea
Down's syndrome

Fig. 4.7

Huntington's chorea is a disease of the nervous system. It is caused by a faulty gene with a dominant allele.

h = normal

H = disease

		Dad has disease	
		h	H
Mum is normal	h	hh	Hh
	h	hh	Hh

2 out of 4 get the disease

Cystic fibrosis is a disease that affects the lungs and digestive system. It is caused by a gene with a recessive allele.

C = normal

c = disease

		Dad has disease	
		c	c
Mum is normal	C	Cc	Cc
	C	Cc	Cc

None get the disease

Mum and Dad are normal but carry the recessive allele.

		Dad is a carrier	
		C	c
Mum is a carrier	C	CC	Cc
	c	Cc	cc

1 out of 4 get the disease

Down's syndrome occurs when the ovum that is fertilised has an extra chromosome. This happens because at meiosis, the chromosomes do not divide properly. One ovum gets 22 and is infertile. The other gets 24. This extra set of genetic instructions usually results in some degree of mental and physical disability.

The baby will have 47 chromosomes instead of the usual number of 46.

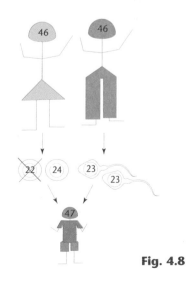

Fig. 4.8

The gene

DNA is the chemical language of life. Unlike English which has 26 letters, the language of DNA has 4 chemical 'letters'. These 'letters' are:

- **A**denine

- **T**hymine

- **G**uanine

- **C**ytosine.

> **KEY POINT** These chemicals are called bases.

Tens of thousands of these bases are arranged along the DNA and spell out the instructions for making proteins.

... ATTGCACTGACTGCATAAGTGTCAACACTCGAG ...

To interpret this language we need to know that three bases are the code for one amino acid... and amino acids join together to make protein.

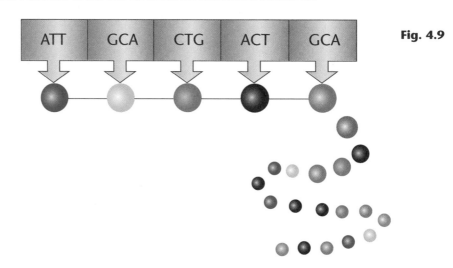

Fig. 4.9

A gene is all the bases on the DNA that code for one protein.

DNA the double helix

DNA actually consists of two strands and is coiled into a double helix. The strands are linked together by a series of paired bases.

A always pairs with T.
G always pairs with C.

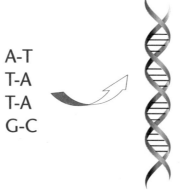

A-T
T-A
T-A
G-C

Fig. 4.10

This is called the Human Genome Project.

Scientists around the world have finally completed the decoding of all the bases in the human DNA. This means we now have a genetic map of all the human genes. This is an amazing and historic achievement and will enable massive advances in medical care and genetic engineering.

Cloning, selective breeding and genetic engineering

AQA
Edexcel A Edexcel B
OCR A ᴬ OCR A ᴮ
NICCEA
WJEC

> **KEY POINT** Clones are genetically identical to each other.

Some plants reproduce by asexual reproduction. The offspring are all genetically identical to their parents. Examples include:

Fig. 4.11

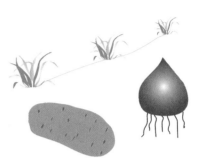

- **runners**, e.g. a strawberry plant

- **bulbs**, e.g. a daffodil

- **tubers**, e.g. a potato plant.

> **KEY POINT** These plants all produce clones.

Commercial growers use cloning by **taking cuttings** of their plants. It is a quick way to increase the numbers of plants for sale.

A more modern way of cloning plants very quickly is to use **micropropagation**. This includes:

- **tissue culture** – using a small group of cells from part of a plant

 Fig. 4.12

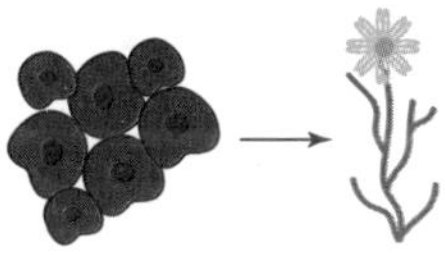

- **embryo transplants** – splitting apart cells from an embryo and transplanting the new embryos into different host mothers

Fig. 4.13

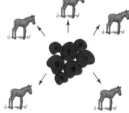

- **nuclear transplant** – replacing the nucleus of an ovum with a nucleus from another cell, e.g. 'Dolly' the sheep.

Fig. 4.14

new nucleus added   old nucleus removed

Selective breeding

This involves selecting those individuals to breed who have the desired characteristics. Farmers have done this for thousands of years.

Fig. 4.15

For example to produce larger hens' eggs, farmers bred the hens that produced the largest eggs, with cocks hatched from large egg laying  mothers. They repeated this for several generations. However, eggs cannot get bigger for ever. Once a hen has all of the 'big egg' alleles, that is as big as it gets.

Other examples include: breeds of dogs, higher yielding crops with better flavour and resistance to disease.

Eugenics Some people thought that this might be a good idea to try with humans.

Genocide Adolf Hitler and his followers believed in an Aryan master-race. He decided to kill what they regarded as inferior races and mentally and physically handicapped people of their 'own race'.

Genetic Engineering

All living organisms use the same language of DNA. The four letters **A**, **G**, **C** and **T** are the same in all living things. Thus a gene from one organism can be removed and placed in a totally different organism where it will continue to carry out its function.

> This means the DNA of a frog would be understood by the DNA of a daffodil.

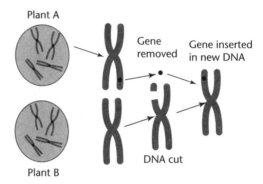

Fig. 4.16

What do you think?

● Some people think that genetic engineering is against 'God and Nature' and is potentially dangerous.

● Some people think that genetic engineering will provide massive benefits to mankind, like better food and less disease.

PROGRESS CHECK

1. Who do boys inherit their Y chromosome from?
2. Explain Mendel's 'Law of Segregation'.
3. State the genotype of a non-tongue roller.
4. If dad is a homozygous tongue roller and mum is a homozygous non-tongue roller, what proportion of their children will be homozygous?
5. If mum and dad are heterozygous, what proportion of their children will be non-tongue rollers?
6. State why there are no 'carriers' for the disease Huntington's chorea.
7. State why both mum and dad have to be carriers, to produce a child with cystic fibrosis.
8. A Down's syndrome child has 47 chromosomes. State why the ovum with 22 chromosomes is rarely fertilised.
9. State how many bases code for one amino acid.
10. State the name of the section of DNA that codes for one protein.
11. State three different ways of performing micropropagation.
12. Explain how selective breeding could be used to increase a cattle herd's milk yield.
13. Genetic engineering. Good or bad – what do you think?

1. Dad; 2. Alleles of a gene separate into different gametes; 3. tt; 4. None; 5. 1 in 4 (3:1); 6. It is a dominant gene and anyone with the gene will have the disease; 7. It is recessive and the child must inherit a recessive allele from both parents; 8. One chromosome is missing making it infertile; 9. Three; 10. Gene; 11. Tissue culture, embryo transplants, nuclear transplants; 12. Breed from high yield cows and bulls who produced high yield cows.

4.3 Evolution

After studying this section you should be able to:

- **explain the evidence for evolution**
- **understand the mechanism of evolution**
- **understand what is meant by extinction.**

Evidence for evolution

AQA
Edexcel A   Edexcel B
OCR A ᴬ   OCR A ᴮ
NICCEA
WJEC

Fossils provide most of the evidence. They tell us about organisms that lived millions of years ago. They can be dated and show how organisms have changed over time.

The evidence is circumstantial. It is not proof. This is why it is called **Darwin's Theory of Evolution** and not Darwin's Law.

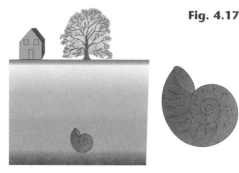

Fig. 4.17

Mechanism of evolution

Charles Darwin (1809–1882) was a naturalist on board HMS Beagle. His job was to make a record of the wildlife seen at the places they visited.

Darwin noticed four things:

- organisms produce more offspring than they need to replace themselves
- population numbers usually remain constant over long time periods
- sexual reproduction produces **variation**
- these variable characteristics are inherited from their parents.

From these four facts, Darwin produced his **Theory of Evolution by Natural Selection**.

If we apply Darwin's Theory to the 'peppered moth' it goes something like this...

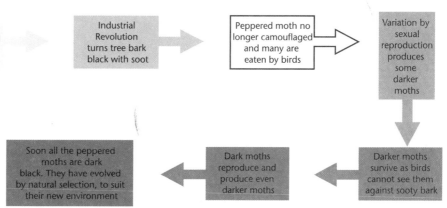

Fig. 4.18

We can also use the Theory to explain how bacteria become resistant to antibiotics.

It took many years before Darwin's Theory was generally accepted. People believed that God made man and they were not prepared to believe that humans had evolved like all other life on Earth.

Wrong!: A man called Lamarck thought that organisms just grew and changed to fit the environment. He thought that the giraffe had a long neck because it just grew so the giraffe could eat leaves from the tallest tree.

Extinction

This is probably what happened to the dinosaurs.

When a species cannot evolve fast enough to compete in a changing environment, it may become extinct. This is more likely to happen if the environment changes very quickly, such as when a major catastrophic climate change takes place, e.g. when an asteroid hits the Earth.

PROGRESS CHECK

1. Explain why Darwin's Theory is not Darwin's Law.
2. Explain why fossils can tell us about how evolution might have happened.
3. State the four observations that Darwin made about natural populations.
4. Use Darwin's Theory to explain how bacteria become resistant to antibiotics.
5. Explain why Darwin's Theory took a long time to be accepted.
6. State the processes that lead to the extinction of a species.

1. It has never been proved, as the evidence although almost universally accepted, is only circumstantial; 2. Fossils are the preserved remains of ancient animals and plants. Their ages can be dated so that they show the sequence of changes that occurred to the organisms; 3. Organisms produce more offspring than is needed to replace them. The population usually remains constant over long time periods. Sexual reproduction produces variation. This variation is inherited from their parents; 4. Antibiotics are used to kill bacteria. A small number of bacteria are more resistant to the antibiotic than the rest because of variation. These bacteria survive and multiply producing many resistant bacteria. Non-resistant bacteria are killed leaving only resistant ones. 5. Most people believed that God created life and could not accept that happened by itself. People also found it difficult to accept that humans could have evolved from a supposedly 'lower' form of life'; 6. Rapid and large environmental changes. Not enough variation in the offspring to survive the changes. All die. Species extinct.

Sample GCSE question

1.

(a) John could roll his tongue, but his sister, Jane could not. John's mother could roll her tongue but his father could not. Complete the following genetic diagram, to show the genotypes of John's family. **[4]**

		mum	
		T	t
d a d		John	
d a d			Jane

		mum	
		T	t
d a d	t	John Tt	
d a d	t		Jane tt

John could not have got T from his dad and Jane must have got t from her mum. Therefore mum must be Tt.

Children always get one allele from each parent.

(b) John knows that tongue rolling is inherited. He also knows that intelligence is a mixture of inherited and environmental factors.

Explain how studies using identical twins can be used to determine how much each contributes to a person's intelligence. **[3]**

Twins have identical genes ✓. If they are separated at birth they each experience different environmental factors ✓. Any difference in intelligence therefore, must be due to these factors ✓.

Twin studies can be used to determine the 'nature or nurture' argument for many other factors.

(c) Some diseases can be caused by faulty genes or chromosomes. Down's syndrome is caused by having an extra chromosome. Complete the following diagram to show how many chromosomes should be present at each stage. **[4]**

When gametes are produced they should receive one chromosome from each pair.

The ovum that contains 22 chromosomes will be infertile because genetic information will be missing.

(d) Diabetes is a disease that is caused by lack of the hormone insulin. Explain how genetic engineering can be used to create bacteria that can make human insulin. **[6]**

The insulin gene is removed from a human chromosome (DNA) ✓. It is cut out using an enzyme ✓. The same enzyme is used to open a bacterial plasmid (DNA) ✓. The human gene is then inserted into the bacterial plasmid ✓. Another enzyme is used to stick the plasmid back together ✓. The bacteria then starts making human insulin ✓.

DNA is a language that all living organisms understand.

Yeast can also be used instead of bacteria as a host for the human gene.

Exam practice questions

1. Susan and Jackie were identical twins. Susan had dyed her hair brown and had a sun tan. Jackie had dyed her hair blonde and did not have a sun tan.

(a) State one characteristic that both twins would have inherited and one characteristic that was due to the environment. [2]

Susan and Jackie also had a sister called Jane. She was not an identical twin.

(b) Explain why Jane did not look like her two sisters. [2]

(c) Jane suffers from a disease called cystic fibrosis. Her sisters do not.

It is caused by a recessive allele.

(i) Complete the checkerboard to show how she inherited this disease. [2]

		mum	
		C	c
d a d	C	w	x
		....	....
	c	Y	Z
		....	....

(ii) State which of the boxes, w, x, y or z, represents Jane. [1]

(iii) State what proportion of the children you would expect to have the disease. [1]

(iv) State what word is used to describe a person who does not have the disease, but can pass it on to the next generation. [1]

(v) State the genotype of the child who will not be able to pass the disease on to the next generation. [1]

(vi) State the possible genotypes of Jackie. [2]

Susan decides that when she is older, she would like to have children of her own. She has a test. Her doctor tells her that her genotype is Cc and that she may pass the disease on to her children. Her doctors tell her that by the time she has children of her own, it will be possible to select embryos that are perfectly normal.

(d) Discuss the moral implications of being able to select desirable genetic qualities for our future children.

2. Richard was feeling ill. His doctor gave him some antibiotics. After several days Richard did not feel any better.

(a) His doctor said that the bacteria were probably drug-resistant.

(i) Suggest what his doctor meant by drug-resistant. [1]

(ii) His doctor gave him a combination of two different antibiotics.

Explain why this is more likely to be an effective treatment. [2]

Exam practice questions

The diagram shows some bacteria, similar to the ones that attacked Richard.

The dark coloured bacteria are a new 'super bug' resistant to all known antibiotics.

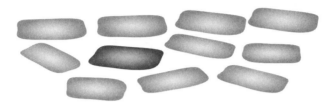

(b) Explain why the new 'super bug' is likely to increase in numbers until all bacteria are antibiotic resistant. **[3]**

Charles Darwin suggested that organisms evolve because:

- they produce more offspring than they need to replace themselves
- sexual reproduction ensures that all the offspring are slightly different.

(c) Explain how these two facts allow organisms to evolve. **[4]**

3. DNA consists of four different chemical bases forming a double helix.

(a) Complete the missing bases to show the structure of this double helix.

C —— ………

G —— ………

A —— ……… **[3]**

The diagram shows a strand of DNA coding for amino acids.

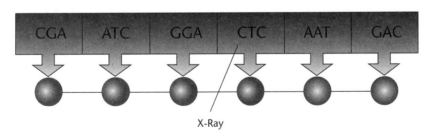

(b) An X-ray damages and removes one of the C bases.

(i) State the name that describes changes of this kind to the structure of DNA. **[1]**

(ii) Explain the effect that the removal of this base will have on the protein that is being produced. **[3]**

(iii) State two other ways by which the DNA can be changed in this way. **[2]**

The following topics are covered in this section:

- Living together
- Energy and nutrient transfer
- Human impact on the environment

What you should know already

Finish the passages using words from the list. You may use the words more than once.

compete	food chain	habitat	oxygen	photosynthesis
primary consumers	producer	respiration	top consumers	trophic level

The diagram shows the contents of an aquarium.

A number of different organisms live in an aquarium. An area where organisms live is called a 1._____ The pond-weed can produce food by 2._____ and so is called a 3._____ The snails eat the pond weed and so are called 4._____. The fish are the 5._____ in this aquarium. Each of these feeding levels is called a 6._____. Listing the organisms in this way, in order to show the passage of food, is called a 7._____

The organisms in this habitat rely on each other for other reasons apart from food. The pond weed produces 8._____ gas by 9._____ that the animals can then use for 10._____. A different type of fish may be introduced into the aquarium. This type of fish may also eat snails and so may 11._____ with the original fish for food.

ANSWERS

1. habitat; 2. photosynthesis; 3. producer; 4. primary consumer; 5. top consumer; 6. trophic level; 7. food chain; 8. oxygen; 9. photosynthesis; 10. respiration; 11. compete

5.1 Living together

After studying this section you should be able to:

● *explain the term competition*
● *realise that the number of organisms in a habitat depends on how much food is available*
● *explain how certain organisms are adapted*
● *understand that certain organisms can closely cooperate with other organisms.*

Competition

AQA
Edexcel A **Edexcel B**
OCR A ᴬ **OCR A** ᴮ
NICCEA
WJEC

Different organisms live in different environments.

> **KEY POINT**
> The place where an organism lives is called its habitat and all the organisms that live there are called the community.

There are always different types of organisms living together in a habitat and many of them are after the same things.

> **KEY POINT**
> This struggle to gain resources is called competition.

> Organisms of the same species are more likely to compete with each other because they have similar needs.

Plants usually compete for:

● light for photosynthesis
● water
● minerals.

Animals usually compete for:

● food to eat
● water to drink
● mates to reproduce with
● space to live in.

Predators and prey

AQA
Edexcel A **Edexcel B**
OCR A ᴬ **OCR A** ᴮ
NICCEA
WJEC

The most common resource that animals compete for is food. Animals obtain their food in a number of different ways.

> All parasites harm their host but a well adapted parasite will not kill the host because it would then need to find another.

| **Parasites** feed off a living organism called the **host** | Food | **Predators** kill and eat other animals called **prey** |

Fig. 5.1

The numbers of predators and prey in a habitat will vary and will affect each other. The size of the two populations can be plotted on a graph that is usually called a predator–prey graph.

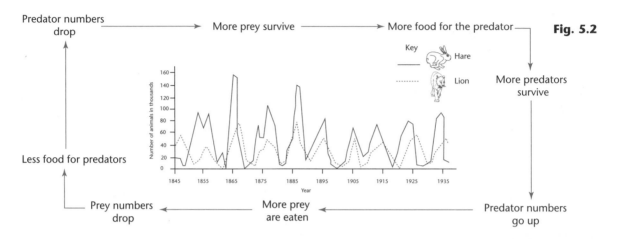

Fig. 5.2

Predator numbers drop ⟶ More prey survive ⟶ More food for the predator ⟶ More predators survive

Less food for predators ⟵ Prey numbers drop ⟵ More prey are eaten ⟵ Predator numbers go up

Key: Hare, Lion

Adaptation

Because there is constant competition between organisms, the best suited to living in the habitat survive. Over many generations the organisms have became suited to their environment.

> **KEY POINT** The features that make organisms well suited to their habitat are called adaptations.

The way in which organisms become adapted to their habitat is explained in Chapter 4.

Habitats, such as the arctic ice floes and deserts, are difficult places to live because of the extreme conditions found there. Animals and plants have to be well adapted to survive:

Fig. 5.3

Fig. 5.4

Polar bears have:
- a large body that holds heat
- thick insulating fur
- a thick layer of fat under the skin
- white fur that is a poor radiator of heat and provides camouflage.

Cacti have:
- leaves that are just spines, to reduce surface area
- deep or widespread roots
- water stored in the stem.

Fig. 5.5

Camels have:
- a hump that stores food as fat
- thick fur on top of body for shade
- thin fur on rest of body.

Cooperation

AQA
Edexcel A   Edexcel B
OCR A ^A   OCR A ^B
NICCEA
WJEC

Instead of competing with each other or trying to eat each other, some different types of organisms have decided to work together.

> **KEY POINT** When two organisms of different species work together so that both gain, this is called mutualism.

Certain bacteria and plants also live together, showing mutualism. This relationship is discussed later in this topic.

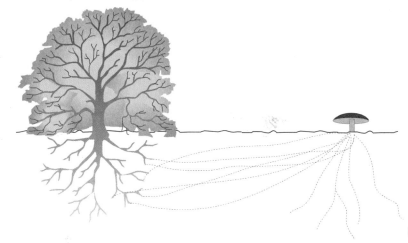

Fig. 5.6

There are many examples of mutualism between different organisms.

Some fungi can live together with trees. The hyphae of the fungi may join with the roots of the tree. The tree passes food to the fungus. The fungus passes water and minerals to the plant.

PROGRESS CHECK

1. A pond contains a large number of stickleback fish. Many pondweed plants are floating on the surface and the sticklebacks feed on this.
   (a) What is the habitat mentioned?
   (b) Write down the name of one population mentioned.
   (c) Write down three things that the sticklebacks are competing for.
2. Some daisy plants are growing under a tree. They are competing with the tree and each other. What are they competing for?
3. An owl has just killed a mouse and is eating it. On the owl's feathers live small mites that regularly eat small amounts of the feathers.
   In this example write down the name of:
   (a) a predator   (b) a prey animal   (c) a parasite   (d) a host.
4. Write down two ways that having spines for leaves help a cactus to survive in the desert.
5. Why does a camel have webbed feet?
6. Small fish often live with larger fish. The small fish feed on small parasites on the scales of the larger fish. Explain why this is an example of mutualism.

1. (a) the pond   (b) pondweed or sticklebacks   (c) oxygen, water, food, mates;   2. Water, minerals, light;   3. (a) owl   (b) mouse   (c) mites   (d) owl;   4. Reduces water loss and protects cactus from animals;   5. To stop it sinking into the sand;   6. The small fish gain food and the large fish gain by having their parasites removed.

5.2 *Human impact on the environment*

LEARNING SUMMARY

After studying this section you should be able to:

- *understand that the human population is increasing and making greater demands on resources*
- *realise that these demands have led to pollution*
- *explain why overexploitation of resources has occurred*
- *explain how conservation schemes have tried to prevent too much damage happening.*

Population size

AQA
Edexcel A   Edexcel B
OCR A ^A   OCR A ^B
NICCEA
WJEC

The number of humans living on Earth has been increasing for a long time but it is going up more rapidly than ever before.

KEY POINT This increase is called a population explosion.

The increasing size of the human population has meant that there has been a greater demand for land.

In many countries, doctors are trying to reduce the population explosion by educating people about contraception.

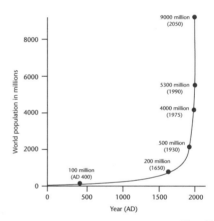

Fig. 5.7

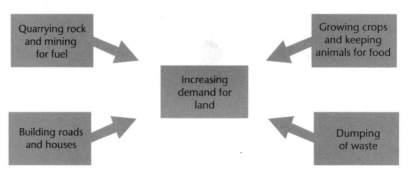

Fig. 5.8

This increased demand for land and resources has meant that many organisms have decreased in numbers. This is because:

Animals have been over-hunted for food

The habitats of many organisms have disappeared

Harmful chemicals have killed organisms

Fig. 5.9

Pollution

AQA
Edexcel A   Edexcel B
OCR A ^A   OCR A ^B
NICCEA
WJEC

Modern methods of food production and the increasing demand for energy have caused many different types of **pollution**.

> **KEY POINT**
>
> Pollution is the release of substances that harm organisms into the environment.

The table shows some of the main polluting substances that are being released into the environment.

polluting substance	main source	effects on the environment
carbon dioxide	burning fossil fuels	*greenhouse effect*
carbon monoxide	car fumes	reduces oxygen carriage in the blood
fertilisers	intensive farming	*eutrophication*
heavy metals	factory waste	brain damage and death
herbicides	intensive farming	some cause mutations
methane	cattle and rice fields	*greenhouse effect*
sewage	human and farm waste	*eutrophication*
smoke	burning waste and fuel	smog and lung problems
sulphur dioxide	burning fossil fuels	*acid rain*

Fig. 5.10

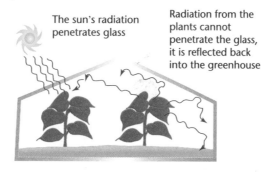

The sun's radiation penetrates glass

Radiation from the plants cannot penetrate the glass, it is reflected back into the greenhouse

The **greenhouse effect** is caused by a build-up of certain gases, such as carbon dioxide and methane, in the atmosphere. These gases trap the heat rays as they are radiated from the earth. This causes the Earth to warm up. This is similar to what happens in a greenhouse.

Fig. 5.11

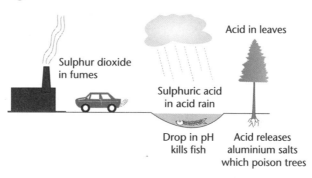

Sulphur dioxide in fumes

Acid in leaves

Sulphuric acid in acid rain

Drop in pH kills fish

Acid releases aluminium salts which poison trees

Acid rain is caused by the burning of coal and oil that contains some sulphur impurities. This gives off sulphur dioxide, which dissolves in rainwater to form sulphuric acid. This falls as acid rain.

> Some organisms need lower amounts of oxygen to survive than others. The variety of organism that is found in a river can be used to tell how polluted the river is.

Eutrophication is caused by sewage or fertilisers being washed into rivers or lakes. The fertilisers cause algae to grow in the water. In the winter most of these die. The dead algae or the sewage is fed on by bacteria that use up all the oxygen in the water. This causes all the other organisms in the water to die.

Over-exploitation

AQA
Edexcel A   Edexcel B
OCR A ^A   OCR A ^B
NICCEA
WJEC

As well as causing pollution, the increasing demands for food and land have caused people to cut down large areas of forests. Some animals have been hunted, until their numbers have been dramatically reduced.

KEY POINT Taking too many natural resources out of the environment is called over-exploitation.

Some natural resources are called non-renewable because they are replaced at such a slow rate. Examples of these are fossil fuels. Many people think that we should switch to other renewable sources of energy.

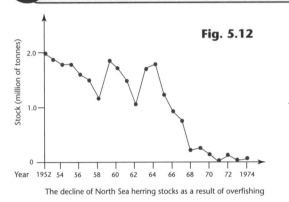

Fig. 5.12

The decline of North Sea herring stocks as a result of overfishing

The European herring was overfished in the 1950s and 1960s. By 1974, the number of surviving herrings was very low.

Other animals have not been so lucky. They have been hunted until no more exist. They are extinct. An example is the woolly mammoth.

Fig. 5.13

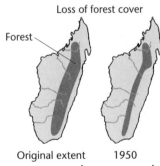

Loss of forest cover

Forest

Original extent   1950

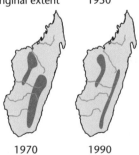

1970   1990

Large areas of tropical rainforest are being cut down. The wood is used as timber or just burnt. The land is used for building houses and roads or for farming.

This is called **deforestation** and is having several effects:

● the rainforests are home to many rare organisms and they are losing their habitat

● the loss and the burning of the trees is making the greenhouse effect worse

● the soil is not held together any more and is being eroded.

Madagascar is a large island near Africa. It contains many rare animals in tropical rainforests. In the last 70 years it has lost 80% of the forests.

Fig. 5.14

Conservation

AQA
Edexcel A Edexcel B
OCR A ᴬ OCR A ᴮ
NICCEA
WJEC

Many people believe that is wrong for humans to damage natural habitats and cause the death of animals and plants. There are many reasons given, such as:

- losing organisms may have unexpected effects on the environment, such as the erosion caused by deforestation

- people enjoy seeing different animals and plants

- some organisms may prove to be useful in the future, for breeding, producing drugs or for their genes

- humans do not have a right to wipe out other species.

> **KEY POINT** Many people are trying to preserve habitats and keep all species of organisms alive. This is called conservation.

Conservation can be helped by adopting the 'three R's': *Reduce, Re-use and Recycle.*

To be able to save habitats and organisms, people must find methods of meeting the ever increasing demand for food and energy, without causing pollution or over-exploitation.

> **KEY POINT** This environmentally friendly growth is called sustainable development.

In 1992, over 150 nations attended a meeting in Brazil called the Earth Summit. They agreed on ways in which countries could work together to achieve sustainable development.

Fig. 5.15

They agreed to:

- reduce pollution from chemicals such as carbon dioxide. This can be done by cutting down on the waste of energy or by using sources of energy that do not produce carbon dioxide

- reduce hunting of certain animals, such as whales.

The document that they signed was called Agenda 21 and local governments are being encouraged to set up local schemes to help with conservation. This is called the **Local Agenda 21**.

 PROGRESS CHECK

1. What is a rapid increase in the size of a population called?
2. Write down three things that an increasing population needs land for.
3. Write down the name of each of these polluting substances:
   (a) A chemical added to crops to supply minerals.
   (b) A chemical that causes acid rain.
   (c) A gas given off by burning fuels that may cause the greenhouse effect.
4. When a pond suffers from eutrophication, why do most of the organisms die?
5. Name an animal that man has hunted to extinction.
6. Write down two reasons why rainforests are being cut down and two effects that this might have on the environment.
7. How does recycling glass bottles help to save energy?

1. A population explosion; 2. Quarrying, building, farming or dumping waste; 3. (a) Fertilisers (b) Sulphur dioxide; (c) Carbon dioxide; 4. They lack oxygen; 5. Woolly mammoth, etc; 6. Room to farm, for timber, to build houses, roads; this may lead to soil erosion; 7. Reduce the amounts of energy and raw materials used to make new glass.

5.3 **Energy and nutrient transfer**

After studying this section you should be able to:

 **LEARNING SUMMARY**

- explain how energy is passed along food chains and is lost all along the chain
- understand how studying this energy flow can help farmers produce more food
- realise that bacteria and fungi carry out an important job in decaying dead material
- understand how this decay allows minerals to be recycled in nature.

Energy transfer

AQA

Edexcel A   Edexcel B

OCR A ᴬ   OCR A ᴮ

NICCEA

WJEC

A food chain shows how food passes through a community of organisms. It enters the food chain as sunlight and is trapped by the producers. They use photosynthesis to trap the energy in chemicals, such as sugars. The energy then passes from organism to organism as they eat each other.

All the organisms in a food chain therefore rely on producers to trap the energy from the Sun.

The mass of all the organisms at each step of the food chain can be measured. This can be used to draw a diagram that is similar to a pyramid of numbers. The difference is that the area of each box represents the mass of all the organisms not the number.

Foxes
Rabbits
Grass

KEY POINT   This type of diagram is called a **pyramid of biomass**.

A pyramid of biomass has some advantages and disadvantages over a pyramid of numbers:

Advantages:

- takes into account the size of each organism, so it is a pyramid.

Disadvantages:

- harder to measure the mass of organisms than to count them
- to measure biomass properly, the organism has to be killed and dried out.

The energy that is lost from a food chain in waste materials is not all wasted. Decomposers will use the waste and start a new food chain.

The reason that a pyramid of biomass is shaped like a pyramid is that energy is lost from the food chain as the food is passed along. This loss of energy happens in a number of ways:

- the organisms do not eat the entire organism that they are feeding on
- they all release energy in their waste products
- energy is lost as heat from respiration.

This means that it needs a larger mass of organisms at one feeding level to support the next level along the chain. This explains the shape of a pyramid of numbers.

Food production

By studying the flow of energy through food chains, farmers can increase the efficiency of their food production methods.

> **KEY POINT** Using scientific knowledge to get maximum food production is called **intensive farming**.

This food chain shows the energy transfers involved in producing chicken for people to eat.

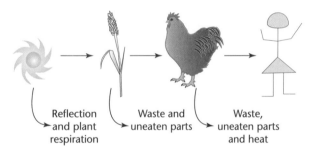

Reflection and plant respiration · Waste and uneaten parts · Waste, uneaten parts and heat

Fig. 5.16

The use of fertilisers can cause pollution in streams and rivers as explained in section 4.2.

In intensive farming the corn is grown using fertilisers and insecticides.

> **KEY POINT** Fertilisers supply plants with minerals for growth and insecticides kill insect pests that feed on the crops.

Using chemical insecticides to kill pests can also cause pollution. Farmers may use other methods, such as using living organisms to control the pests. This is called biological control.

The chickens are kept in warm conditions indoors and do not have to look for their food. They will therefore lose less energy. This will mean more energy is available for humans when they eat the chicken.

Some people do not like the way that animals are kept in intensive farming.
They think it is cruel to keep animals indoors with little space to move.

Vegetarians do not eat meat. They say that by eating plant material, the energy has to pass through one less step. This means less is wasted.

INTENSIVE FARMING METHODS ARE NOT LIKED BY EVERYBODY

Other people prefer to eat plants that have been produced organically. This does not involve the use of chemical fertilisers or insecticides.

Fig. 5.17

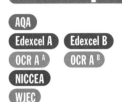
Decomposers

AQA
Edexcel A   Edexcel B
OCR A ᴬ   OCR A ᴮ
NICCEA
WJEC

Some animals and plants die before they are eaten. They also produce large amounts of waste products. This waste material must be broken down because it contains useful minerals. If this did not happen, organisms would run out of minerals.

> **KEY POINT** Organisms that break down dead animal and plant material are called **decomposers.**

Decomposers will also break down material that humans are storing for food. Different methods of preserving food are used to stop the decomposers spoiling it.

The main organisms that act as decomposers are bacteria and fungi. They release enzymes on to the dead material that digest the large molecules. They then take up the soluble chemicals that are produced. The bacteria and fungi use the chemicals in respiration and for raw materials.

Gardeners try to provide ideal conditions for decomposers to work in compost heaps.

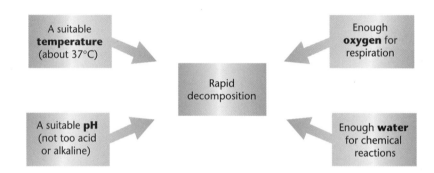

A suitable **temperature** (about 37°C)

Enough **oxygen** for respiration

Rapid decomposition

A suitable **pH** (not too acid or alkaline)

Enough **water** for chemical reactions

Fig. 5.18

Nutrient cycles

AQA
Edexcel A   Edexcel B
OCR A ᴬ   OCR A ᴮ
NICCEA
WJEC

It is possible to follow the way in which each mineral element passes through living organisms and becomes available again for use. Scientists use nutrient cycles to show how these minerals are recycled in nature.

The normal level of carbon dioxide in the air is between 0.03 and 0.04%. This is continuing to rise.

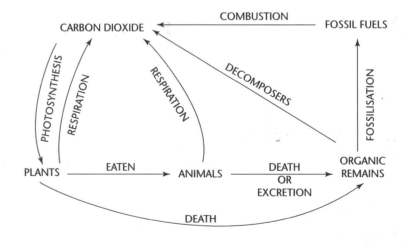

CARBON DIOXIDE   COMBUSTION   FOSSIL FUELS

PHOTOSYNTHESIS   RESPIRATION   RESPIRATION   DECOMPOSERS   FOSSILISATION

PLANTS   EATEN   ANIMALS   DEATH OR EXCRETION   ORGANIC REMAINS

DEATH

Fig. 5.19

The nitrogen cycle is more complicated because as well as the decomposers, it involves three other types of bacteria:

- **nitrifying bacteria** – these bacteria live in the soil and convert ammonium compounds to nitrates. They need oxygen to do this

- **denitrifying bacteria** – these bacteria in the soil are the enemy of farmers. They turn nitrates into nitrogen gas. They do not need oxygen

- **nitrogen fixing bacteria** – they live in the soil or in special bumps called nodules on the roots of plants from the pea and bean family. They take nitrogen gas and convert it back to useful nitrogen compounds.

The nitrogen fixing bacteria and pea plants have a mutualistic relationship. The bacteria are provided with some food from the plant and they fix nitrogen for the plant to use.

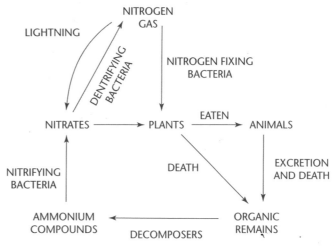

Fig. 5.20

1. How does energy enter a food chain?
2. Why might a scientist chose to construct a pyramid of numbers rather than a pyramid of biomass?
3. Write down three ways that energy is lost from a food chain.
4. Why does keeping chickens indoors mean that they lose less energy?
5. Why does it make sense in terms of energy loss for humans to be vegetarians?
6. Write down the names of two types of decomposer.
7. Why can't decomposers break down food when it is pickled in vinegar?
8. Why is it important that farmers make sure that the soil in their fields contains enough oxygen?

PROGRESS CHECK

1. As sunlight, which is used in photosynthesis;   2. It is easier to count numbers and does not involve killing the organisms;   3. Heat, waste materials and uneaten parts;   4. They do not need to use so much energy to keep themselves warm;   5. The food has to pass through fewer transfers so less energy is lost;   6. Bacteria and fungi;   7. pH is too low;   8. So that nitrifying bacteria can work and make nitrates.

Sample GCSE question

1. Red spider is a pest of plants that grow in greenhouses, such as tomatoes. The spiders can be killed by spraying with an insecticide.

   Another way of killing the spiders is to buy some mites that can be released into the greenhouse. They breed faster than the red spider and eat the red spider.

(a) A gardener used a chemical insecticide to kill the red spider.

   (i) The gardener found that the insecticide killed the spider but when he started using it fewer of his tomato flowers produced tomatoes. Suggest why this is so. [2]

   The insecticide killed pollinating insects ✓. Without pollination of the flowers the tomato fruits cannot develop ✓.

   This is a particular problem when using chemical insecticides in closed areas.

   (ii) Over a number of years it became less effective in killing the spiders. Suggest why this might be so. [1]

   The insect population has become resistant to the insecticide ✓.

   Do not use the word 'immune'. It is not the same as resistant.

(b) The use of the mite is an example of a different type of pest control. What is this called? [1]

   Biological control ✓.

(c) Explain why scientists have to be careful when they introduce this type of control. [2]

   The organism that has been introduced might become a pest itself ✓. It may start to feed on other organisms when it has eaten all of the red spiders ✓.

   There have been a number of occasions when biological control has gone wrong.

Exam practice questions

1. The diagram shows the amount of energy in the cereal food eaten by a cow in a certain time. It also shows the amount of the energy that is used to make new tissue.

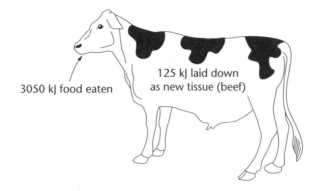

3050 kJ food eaten

125 kJ laid down as new tissue (beef)

(a) **(i)** What percentage of the energy taken in by the cow is trapped in new tissue? [2]

(ii) What happens to the rest of the energy? [2]

(b) **(i)** Explain why in terms of energy capture, it is more efficient for a person to eat the cereal that the cow is eating, rather than beef from the cow. [3]

(ii) Give one reason why people choose to eat beef rather than the cereal. [1]

(c) Write down one way in which a farmer could try to make the production of beef more energy efficient. [1]

2. Organisms are adapted to the environment that they live in.

Explain how each of the following characteristics helps the organism survive.

(a) Camels store large amounts of fat in their humps. [2]

(b) Some cacti have deep roots that pass straight down whereas other types of cacti have shallow roots that spread out a long distance [3]

(c) Polar bears are large animals with very small ears for the size of their body. [2]

(d) The larvae of many insects do not feed on the same type of food as the adult insect. [1]

Extension material

Check in the table on pages 4 and 5 to see which sections you need to study.

Chapter	Section	Studied in class	Revised	Practice questions
6 Classification	Principles of classification			
	The five kingdoms			
	Viruses			
7 Adaptation	Living in water			
	Living on land			
8 Microbes and food	Growth of microbes			
	Preserving food			
	Traditional uses of microbes			
	Modern uses of microbes			
9 Microbes, waste and fuel	Sewage treatment			
	Fuels			
	Other industrial uses of microbes			
10 Microbes and disease	Causes of disease			
	Spread of disease			
	Antiseptics, antibiotics and painkillers			
	Use of antibiotics			
	Vaccines and their production			
11 Genetics and genetic engineering	Structure of DNA			
	Protein synthesis			
	Mutations			
	Gene detection			
	Genetic engineering			
	Genetic fingerprints			
	Cloning			
12 Further physiology	Food and feeding			
	Excretion			
	Coordination and movement			
13 Food production	World food shortage			
	Agriculture			
	Plant disease and its control			
	Hormones and food production			
14 Further ecology	Ecosystems			

6 Classification

The following topics are covered in this section:
- **Principles of classification**
- **The five kingdoms**
- **Viruses**

LEARNING SUMMARY

After studying this section you should be able to:

- explain the principles of a modern classification system
- describe how organisms are scientifically named
- understand the five kingdom classification system
- describe the characteristics of the five kingdoms and realise that it is sometimes difficult to decide on the difference between plants and animals
- describe the structure of viruses
- explain how viruses reproduce.

KEY POINT

Scientists have carried out the process of classifying organisms into groups for a long time. This chapter looks at the modern systems of classification and sees that these methods, as well as being convenient, can tell scientists a lot about the evolution of the organisms.

Principles of classification

OCR A ^A
NICCEA

Artificial versus natural classification

Humans have been classifying organisms into groups ever since they started studying them. This makes it much more convenient when trying to identify an unknown organism.

KEY POINT

The process of placing organisms into groups is called taxonomy.

> This system means that very different organisms end up in the same group.

The system that was used for a long time used one characteristic to classify a group of organisms. For example:

All animals that fly   All animals that swim

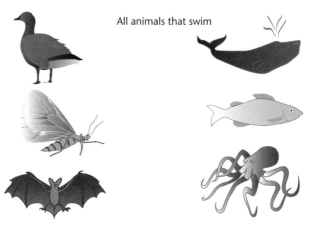

Fig. 6.1

This type of system is called an artificial system.

In the 17th and 18th centuries John Ray and Carl Linnaeus developed a new system. This puts organisms that share the most common characteristics together in groups.

> **The advantage of this type of system is organisms that are in the same group are more likely to have evolved from the same ancestor.**

> **KEY POINT** This type of system is used today and is called a Natural System. The old type of system was an artificial system.

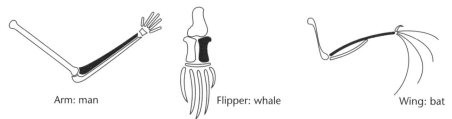

Fig. 6.2

A natural system classifies man, whales and bats in the same group because although their limbs do different jobs, they are all similar in structure and the animals share other similarities.

Grouping and naming

Linnaeus introduced a system of smaller and smaller groups into which organisms are placed. Kingdoms are the largest groups and species the smallest groups. The smaller the groups, the more similar the organisms. For example, a lion is classified as follows:

> **Remember:**
> **P**eas
> **C**arrots
> **O**nions
> **F**or
> **G**ood
> **S**oup

Kingdom	Animalia	–	animals
Phylum	Vertebrata	–	animals with backbones
Class	Mammalia	–	warmblooded animals with hair or fur, suckle young
Order	Carnivora	–	flesh eating mammals, e.g. cats, dogs, bears and seals
Family	Felidae	–	cats, large and small
Genus	*Panthera*	–	certain large cats, such as tigers and panthers
Species	*Leo*	–	lion only

To decide if organisms are similar enough to be in the same species seems obvious but some members of the same species look quite different.

> **KEY POINT** Organisms are put into the same species if they can breed with each other to produce fertile offspring.

Naming organisms

Linnaeus also introduced a universal system for naming organisms. This saved a lot of confusion because organisms had previously had different names in different countries or areas.

KEY POINT **The system Linnaeus introduced is called the binomial system.**

Each organism has two names in Latin, the first is the name of the genus and the second the species, for example:

The binomial name is always in italics when typed. The genus starts with a capital letter but the species does not.

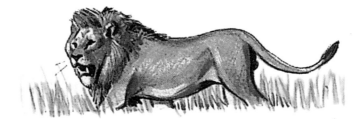

Fig. 6.3

PROGRESS CHECK

1. Which is the correct binomial name for man:
   (a) *Human being* (b) *homo sapiens* (c) *Homo Sapiens* (d) *Homo sapiens*
2. Why do some members of a species look different compared to others?
3. A horse and a donkey can breed to produce an infertile animal called a mule. Are horses and donkeys members of the same species?

1. (d); 2. There is variation in all populations; The individuals have become adapted to their environment; 3. No, if they were from the same species, then the offspring would be fertile.

The five kingdoms

AQA | Edexcel A | Edexcel B | OCR A [A] | OCR A [B] | NICCEA | WJEC

How many kingdoms?

When classifying organisms, the first step is to place them in a kingdom. Most modern systems have five kingdoms.

KEY POINT **The five kingdoms are Bacteria, Fungi, Protoctista, Plants and Animals.**

This system was suggested by the scientists Margulis and Schwatz.

The hardest job is often deciding whether an organism is an animal or a plant. The main difference is the way that they feed.

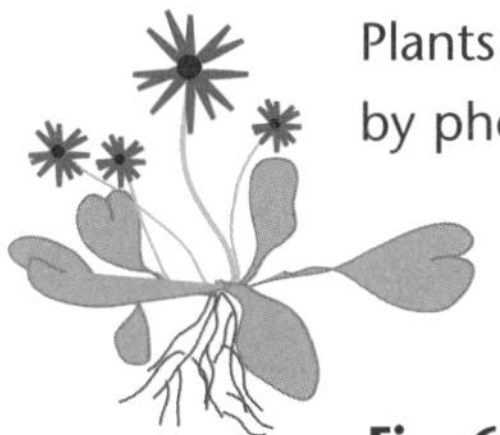

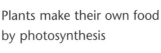

Plants make their own food by photosynthesis

Fig. 6.4

but animals need to take their food in ready-made.

Fig. 6.5

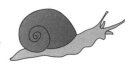

Protoctista

The kingdom protoctista contains simple organisms that used to be classified as plants or animals. This includes the algae and single celled protozoa.

Plants

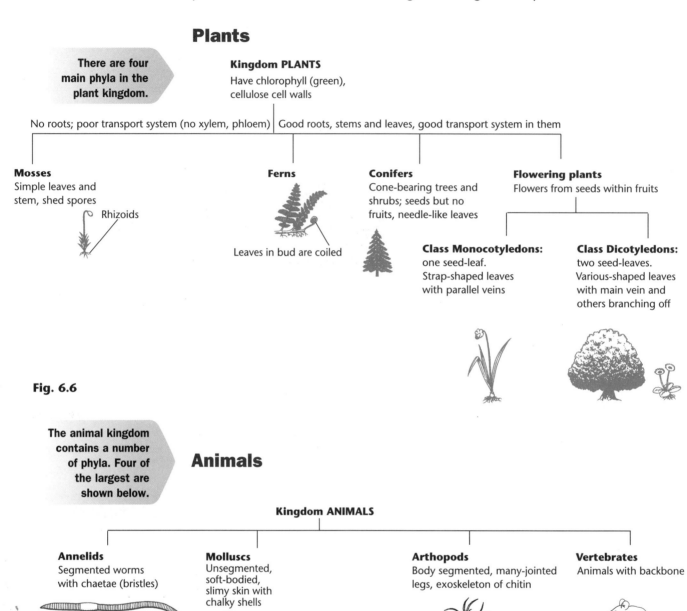

There are four main phyla in the plant kingdom.

Kingdom PLANTS
Have chlorophyll (green),
cellulose cell walls

No roots; poor transport system (no xylem, phloem) | Good roots, stems and leaves, good transport system in them

Mosses
Simple leaves and
stem, shed spores

Rhizoids

Ferns

Leaves in bud are coiled

Conifers
Cone-bearing trees and
shrubs; seeds but no
fruits, needle-like leaves

Flowering plants
Flowers from seeds within fruits

Class Monocotyledons:
one seed-leaf.
Strap-shaped leaves
with parallel veins

Class Dicotyledons:
two seed-leaves.
Various-shaped leaves
with main vein and
others branching off

Fig. 6.6

Animals

The animal kingdom contains a number of phyla. Four of the largest are shown below.

Kingdom ANIMALS

Annelids
Segmented worms
with chaetae (bristles)

Molluscs
Unsegmented,
soft-bodied,
slimy skin with
chalky shells

Snail

Arthopods
Body segmented, many-jointed
legs, exoskeleton of chitin

Spider

Vertebrates
Animals with backbone

Rat

Fig. 6.7

Bacteria

The kingdom containing the simplest organisms includes all Bacteria. They vary in size and shape but are all single cells. They are smaller than animal or plant cells with a simpler internal structure.

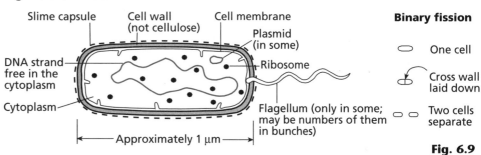

A generalised bacterium

Slime capsule

Cell wall
(not cellulose)

Cell membrane

Plasmid
(in some)

DNA strand
free in the
cytoplasm

Ribosome

Cytoplasm

Flagellum (only in some;
may be numbers of them
in bunches)

← Approximately 1 µm →

Binary fission

One cell

Cross wall
laid down

Two cells
separate

Fig. 6.9

A micrometer (µm) is
0.001 of a mm.
Animal cells are often
30 µm across.

> **KEY POINT** Bacteria contain DNA which is found in the cytoplasm, never contained in a nucleus.

Fungi

The last kingdom contains the Fungi. They share some of the properties of plants but unlike plants, they cannot make their own food.

> **KEY POINT** Fungi are made up of many tubes of cytoplasm called hyphae. They do not contain chlorophyll and although they have cell walls, they are not made of cellulose.

Two common types of fungi are moulds and yeast.

Fig. 6.10

Yeast is unlike most
fungi because it is not
made up of hyphae. It
is made up of
individual cells.

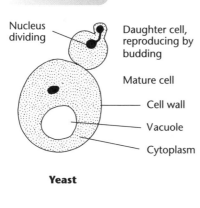

Nucleus
dividing

Daughter cell,
reproducing by
budding

Mature cell

Cell wall

Vacuole

Cytoplasm

Yeast

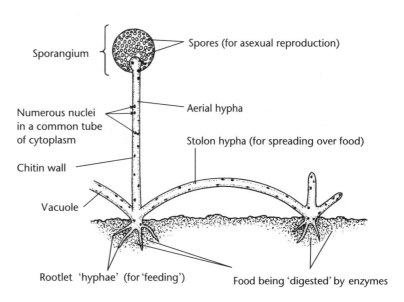

Part of a mycelium from a mould

Sporangium

Spores (for asexual reproduction)

Numerous nuclei
in a common tube
of cytoplasm

Aerial hypha

Stolon hypha (for spreading over food)

Chitin wall

Vacuole

Rootlet 'hyphae' (for 'feeding')

Food being 'digested' by enzymes

> **PROGRESS CHECK**
>
> 1. Name the five kingdoms used in most classification systems.
> 2. What is the main difference between animals and plants?
> 3. Which main characteristics do fungi share with plants and which with animals?
>
> 1. bacteria, protoctista, plants, animals and fungi; 2. The way in which they feed. Plants make their own food whereas animals take it in ready-made. 3. Fungi need to take food in ready-made like animals. They have a cell wall like plants although it is not made of cellulose. Like plants they tend to respond slowly to stimuli and like plants they cannot move.

Viruses

AQA
Edexcel A   Edexcel B
OCR A ᴬ   OCR A ᴮ
NICCEA
WJEC

Viral structure

Viruses are smaller than living cells, even bacteria.

> **KEY POINT**
> They consist of a protein coat, which surrounds a strand of genetic material. They are not cells.

A nanometer (nm) is a thousandth of a micrometer.

The genetic material is DNA in some viruses and RNA in others.

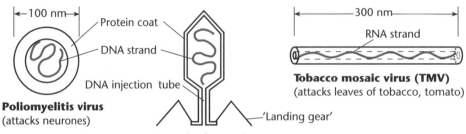

←100 nm→
Protein coat
DNA strand
DNA injection tube
Poliomyelitis virus
(attacks neurones)

Bacteriophage virus
(plentiful in sewage, killing bacteria)

'Landing gear'

←─────300 nm────→
RNA strand
Tobacco mosaic virus (TMV)
(attacks leaves of tobacco, tomato)

Fig. 6.11

Reproduction in viruses

Viruses need living cells in order to reproduce. In doing so they destroy the cell and so they are all parasites.

> **KEY POINT**
> Viruses inject their genetic material into the cell and take over the cell. They then use the cell's material to make new viruses.

The diagram shows how a virus attacks a bacterial cell.

Because viruses spend a lot of time inside the host's cells, this makes it difficult to kill them.

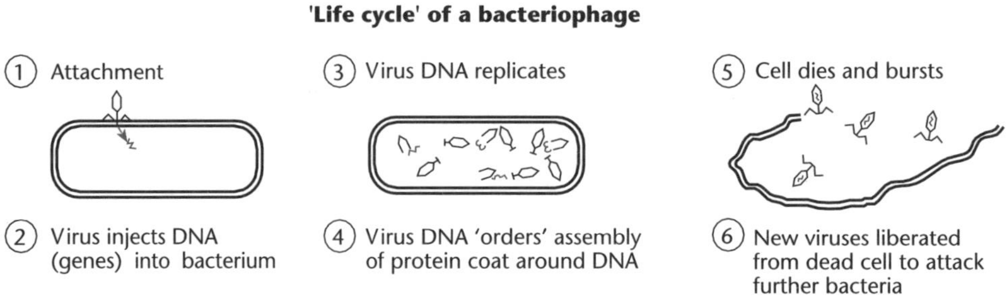

'Life cycle' of a bacteriophage

① Attachment

③ Virus DNA replicates

⑤ Cell dies and bursts

② Virus injects DNA (genes) into bacterium

④ Virus DNA 'orders' assembly of protein coat around DNA

⑥ New viruses liberated from dead cell to attack further bacteria

Fig. 6.12

Although viruses can reproduce inside living cells, they do not grow, feed, excrete or respire and so most people do not really class viruses as living organisms because they do not fulfil all seven characteristics.

PROGRESS CHECK

1. How many times bigger is the generalised bacterium shown on page 93 compared to the poliomyelitis virus shown above?
2. How does a virus take over a living cell?
3. Which characteristics of living organisms do viruses possibly show?

1. 10 times;   2. by injecting its genetic material;   3. They reproduce, and possibly show some movement and sensitivity.

Sample GCSE question

1. The following diagrams show a bacterium and a virus:

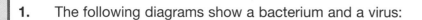

BACTERIUM

VIRUS

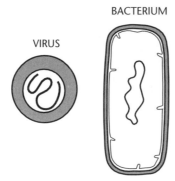

(a) The real length of the bacterium is 0.002 mm.

Work out the magnification of the diagram of the bacterium.

You must show how you worked out your answer. **[2]**

length of bacterial diagram = 40 mm ✓

Magnification = $\dfrac{40}{0.002} = 20000$ ✓

> *In this type of question make sure that you measure the cell in the same units as in the question, i.e. mm.*

(b) Apart from size, write down two differences between the structure of bacteria and viruses. **[2]**

1. Bacteria contain cytoplasm but viruses do not ✓.

2. Bacteria have a cell membrane but viruses do not ✓.

> *These differences are due to the fact that viruses are not cells.*

(c) Explain why bacteria are placed in a kingdom of their own. **[2]**

Bacteria do not have a nucleus, the genetic material is in the cytoplasm ✓.

They are much smaller than plant or animal cells ✓.

> *This is the main characteristic of bacteria.*

(d) Some scientists originally thought that viruses were the ancestors of all living organisms.

Explain why their method of reproduction makes that unlikely. **[2]**

Viruses need living cells in order to reproduce ✓. *If they were the ancestors of living cells they would have to be able to reproduce by themselves* ✓.

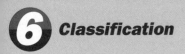

Exam practice questions

1. The diagrams A to E show five different organisms.

A   B   C   D   E

(a) Write down the name of the kingdom in which each organism belongs: **[5]**

A _____

B _____

C _____

D _____

E _____

(b) How does organism C obtain its food?   **[3]**

(c) Which kingdoms are not represented by the organisms shown above?   **[2]**

2. Read the following passage and answer the questions that follow:

> The Greek philosopher Aristotle was the first person to make a real attempt to classify living organisms. Aristotle only knew of several hundred living organisms and the system he devised was an artificial system.
>
> It was not until about 2000 years later that John Ray developed a natural classification system, which was then improved upon by Linnaeus.
>
> Linneaus also devised the binomial system for naming organisms that avoided much confusion.

(a) What is meant by an artificial classification system, such as the one devised by Aristotle?   **[1]**

(b) (i) How does the binomial system of naming organisms work? ⸳⸳⸳⸳⸳⸳⸳⸳⸳⸳⸳⸳⸳⸳⸳⸳ **[2]**

     (ii) How did this naming system 'avoid much confusion'.   **[3]**

(c) Aristotle was only aware that several hundred organisms existed.

Suggest reasons why Ray and Linnaeus were aware of many more organisms.   **[3]**

The following topics are covered in this section:

- *Living in water* - *Living on land*

LEARNING SUMMARY

After studying this section you should be able to:

- describe how fish breathe and move in water
- describe how filter feeders obtain their food
- describe how birds move
- describe a selection of insect feeding mechanisms
- explain how plants prevent water loss and achieve pollination.

KEY POINT

Chapter 4 describes how living organisms become adapted to living in particular habitats. This chapter extends this idea by comparing some of the adaptations needed to live in water with those needed to live on land.

Living in water

(AQA)
(OCR A [A])
(NICCEA)

Breathing in fish

KEY POINT

The respiratory surfaces in fish are the gills. They are divided up into a large number of filaments in order to provide a large surface area for gas exchange.

The gill filaments have a rich blood supply in order to extract oxygen from the water.

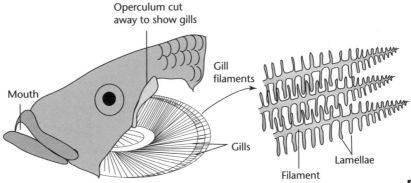

Operculum cut away to show gills

Mouth

Gill filaments

Gills

Filament

Lamellae

Fig. 7.1

Air has about 20% oxygen but water usually has less than 1%.

Remember: polluted or warm water has less oxygen dissolved in it so fish find it harder to obtain oxygen.

There is much lower level of oxygen dissolved in water than in air and so fish must keep a rapid stream of water passing over their gills. The floor of the mouth is lowered, sucking water in. The mouth closes and then the floor is raised, pushing water over the gills and out under the operculum.

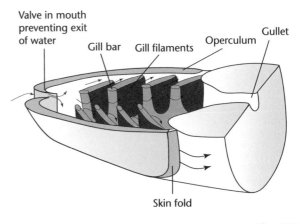

Valve in mouth preventing exit of water

Gill bar

Gill filaments

Operculum

Gullet

Skin fold

Fig. 7.2

Movement in fish

Water is much denser than air so this helps to support fish allowing them to grow large. It does, however make it difficult to move quickly.

> **KEY POINT** In order to overcome the resistance to movement caused by water, fish have a number of adaptations for movement.

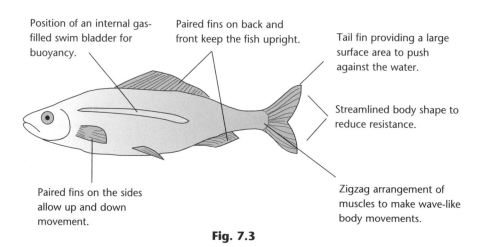

Position of an internal gas-filled swim bladder for buoyancy.

Paired fins on back and front keep the fish upright.

Tail fin providing a large surface area to push against the water.

Streamlined body shape to reduce resistance.

Paired fins on the sides allow up and down movement.

Zigzag arrangement of muscles to make wave-like body movements.

Fig. 7.3

Filter feeding

Many animals that live in water feed on small particles or organisms that are suspended in the water.

> **KEY POINT** Many animals take in a current of water for respiration. This water can also be filtered to obtain the food. This is called filter feeding.

Even large animals, such as certain sharks, are filter feeders.

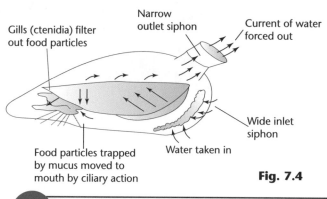

Gills (ctenidia) filter out food particles

Narrow outlet siphon

Current of water forced out

Wide inlet siphon

Water taken in

Food particles trapped by mucus moved to mouth by ciliary action

Fig. 7.4

Small hairs called cilia line the gills and beat to draw water over the gills. The food gets trapped in sticky mucus on the gills and is moved to the mouth by the cilia.

> **KEY POINT** Even large animals, such as certain sharks, are filter feeders

 PROGRESS CHECK

1. Why is it more difficult in water than in air to (a) obtain oxygen and (b) move?
2. How are gill filaments adapted to extract a high proportion of the oxygen from the water?
3. Mussels live on the sea-shore. Why can they not feed at low tide?

1. (a) Water contains much less oxygen than air; (b) Water is more dense than air and so provides more resistance; 2. Large surface area, thin, rich blood supply; 3. At low tide they are not covered with water and so cannot set up a stream of water over their gills.

Living on land

Flight in birds

Air is less dense than water and so this means that birds encounter less resistance to movement than fish. The disadvantage is that air provides less support and less resistance to push against.

> **KEY POINT** Birds have many adaptations that help them to provide the necessary force to fly and to keep their mass to a minimum.

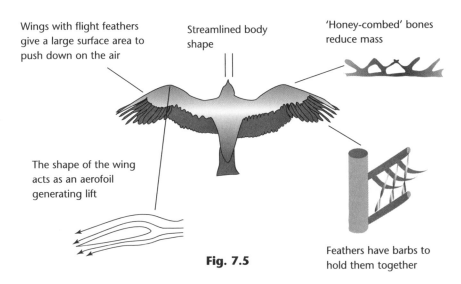

Wings with flight feathers give a large surface area to push down on the air

Streamlined body shape

'Honey-combed' bones reduce mass

The shape of the wing acts as an aerofoil generating lift

Feathers have barbs to hold them together

Fig. 7.5

Reptiles and amphibians

Reptiles and amphibians are different classes of the phylum containing all vertebrates. Many amphibians live on land but they are much more limited than reptiles as to where they can live.

Fig. 7.6

Fig. 7.7

- Amphibians have external fertilisation in which the sperm and eggs are released outside the body.
- The fertilised eggs are covered in jelly and would easily dry out on land.
- The eggs hatch into larvae, such as tadpoles, which have gills and so need to live in water.
- The skin of the adult amphibian is not waterproof.

- Reptiles have internal fertilisation.
- Reptiles' eggs are covered in a waterproof, leathery shell.
- Young reptiles are fully formed miniatures of the adults.
- Reptiles have scaly waterproof skin.

Feeding in insects

Insects are, in terms of numbers, the most successful organisms living on land. There are over 750 000 different species of insects. There are a number of reasons why they are so successful but one is the variety of their feeding methods. They do not have jaws but they have a tremendous variety of mouthparts so that different species feed on different foods.

 KEY POINT The variety of insect feeding methods means that they are not all competing for the same type of food.

 Mosquitoes may inject parasites, such as the malaria parasite, when they feed.

An example of a specialised feeding method is the mosquito's. The mosquito penetrates the skin and injects a substance that stops the blood from clotting.

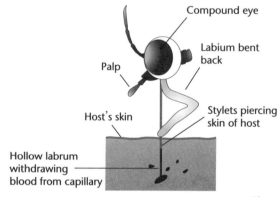

Fig. 7.8

Compound eye
Labium bent back
Palp
Host's skin
Stylets piercing skin of host
Hollow labrum withdrawing blood from capillary

Aphids may also spread pathogens from plant to plant.

Aphids (greenflies) pierce plants to tap the food in the phloem.

Houseflies suck up food that they have partially digested with enzymes and butterflies suck up nectar from flowers.

Some insects undergo a complete change during their life. This means that the adult may feed on a different food from its young. This reduces competition between the adult and the young.

KEY POINT This complete change in body form is called complete metamorphosis.

Fig. 7.9

The adult lays eggs that hatch into larvae.

Larva (e.g. caterpillar)   Pupa (e.g. chrysalis)   Adult (e.g. butterfly)

Plants and living on land

Plants cannot move around looking for water and so living on land presents them with problems. One problem is reproduction. In water, sex cells can swim to each other but on land this is more difficult. Flowering plants have developed pollen grains that contain the male gamete. It is often transferred to another plant by insects.

KEY POINT Flowering plants are usually adapted for pollination by wind or insects.

The mass of the bee on the landing pad moves the anthers or the stigma so that they touch the bee's back.

More variation is produced by cross pollination

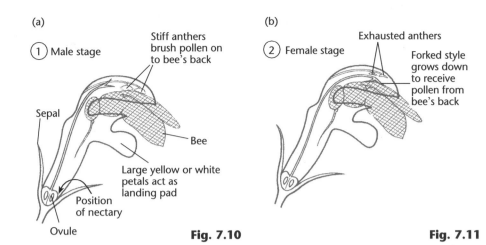

(a)

① Male stage

Stiff anthers brush pollen on to bee's back

Sepal

Bee

Large yellow or white petals act as landing pad

Position of nectary

Ovule

Fig. 7.10

(b)

② Female stage

Exhausted anthers

Forked style grows down to receive pollen from bee's back

Fig. 7.11

The white deadnettle is pollinated by bees. The male parts of the plant ripen before the female parts. This stops the flower from pollinating itself.

Cacti live in some of the driest deserts and so have many adaptations to enable them to gain water and slow down the rate at which it is lost.

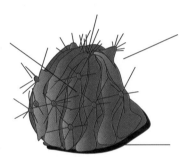

The stem contains chlorophyll to take over the role of the leaves in photosynthesis.

Leaves have become spines, in order to reduce the surface area for water loss.

The stem is swollen with stored water.

Fig. 7.12

PROGRESS CHECK

1. Why do birds have 'hollow' bones?
2. How does a bird gain up-thrust when it flies?
3. What is the difference between reptile and amphibian skin?
4. Why do cacti have spines?

1. To reduce their weight so that it is easier to fly;   2. The wings push down on the air forcing the bird upwards and the shape of the wings acts as an aerofoil;   3. Reptiles have a scaly impermeable skin but amphibians have a smooth permeable skin;   4. The spines are the leaves, reduced in size to limit the surface area for water loss.

Sample GCSE question

1. The following diagram shows part of a bony fish.

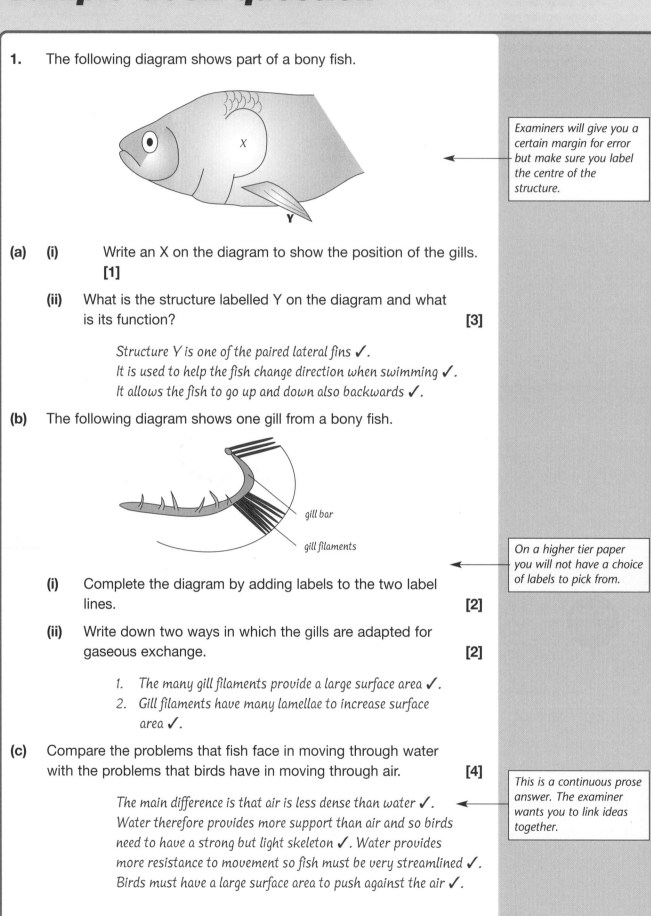

> Examiners will give you a certain margin for error but make sure you label the centre of the structure.

(a) **(i)** Write an X on the diagram to show the position of the gills. **[1]**

(ii) What is the structure labelled Y on the diagram and what is its function? **[3]**

> *Structure Y is one of the paired lateral fins ✓.*
> *It is used to help the fish change direction when swimming ✓.*
> *It allows the fish to go up and down also backwards ✓.*

(b) The following diagram shows one gill from a bony fish.

gill bar

gill filaments

> On a higher tier paper you will not have a choice of labels to pick from.

(i) Complete the diagram by adding labels to the two label lines. **[2]**

(ii) Write down two ways in which the gills are adapted for gaseous exchange. **[2]**

> *1. The many gill filaments provide a large surface area ✓.*
> *2. Gill filaments have many lamellae to increase surface area ✓.*

(c) Compare the problems that fish face in moving through water with the problems that birds have in moving through air. **[4]**

> This is a continuous prose answer. The examiner wants you to link ideas together.

> *The main difference is that air is less dense than water ✓.*
> *Water therefore provides more support than air and so birds need to have a strong but light skeleton ✓. Water provides more resistance to movement so fish must be very streamlined ✓. Birds must have a large surface area to push against the air ✓.*

Exam practice questions

1. The diagram shows two types of cacti that live in the desert.

(a) State two features that both cacti possess that help them to live in dry habitats. **[2]**

(b) Explain how each feature allows the cactus to survive. **[2]**

(c) Many reptiles live in this desert but few amphibians.

Explain why this is. **[3]**

2. The following diagram shows the feeding mouthparts of three different insects.

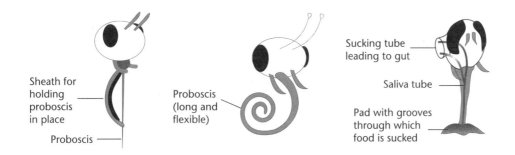

Sheath for holding proboscis in place

Proboscis

Proboscis (long and flexible)

Sucking tube leading to gut

Saliva tube

Pad with grooves through which food is sucked

(a) Complete the following table. **[3]**

insect	example of insect	type of food	brief feeding method
A			
B	butterfly		proboscis uncoils and food is sucked up as though through a straw
C		human food, rotting organic matter	

(b) Two of the insects shown in the diagram may pass on pathogens when they feed.

Describe how this may occur for **one** of these insects. **[2]**

Microbes and food

The following topics are covered in this section:

- Growth of microbes
- Traditional uses of microbes
- Preserving food
- Modern uses of microbes

LEARNING SUMMARY

After studying this section you should be able to:

- recall the conditions needed for microbes to grow
- explain how microbes can be grown on agar plates
- explain why food spoils
- describe various methods used to preserve food
- describe how modern fermenters are used to grow microbes
- describe how microbes have been used in the traditional processes of baking, brewing, preserving dairy produce, making vinegar and soy sauce
- describe how microbes are being used to produce new food sources.

KEY POINT

Microorganisms or microbes is the name given to a range of living organisms that are so small that they can only be seen using a microscope. They include a variety of organisms, such as bacteria, algae, fungi and protozoa.

Many of these organisms are pathogens, they cause disease. However, man has used others for centuries in food production. More recently, new methods of food production have been introduced using other microbes.

Growth of microbes

AQA
Edexcel A   Edexcel B
OCR A ᴬ   OCR A ᴮ
NICCEA

Ideal growth conditions

The main microbes involved in food spoilage and food production are bacteria and fungi.

KEY POINT

These microbes need certain conditions in which to grow at a fast rate.

Remember when scientists talk about 'growth' of microbes, they mean an increase in numbers or reproduction. Individual microbes do not usually increase in size.

In ideal conditions a bacterium can reproduce every twenty minutes producing 4000 million, million, million in 24 hours!

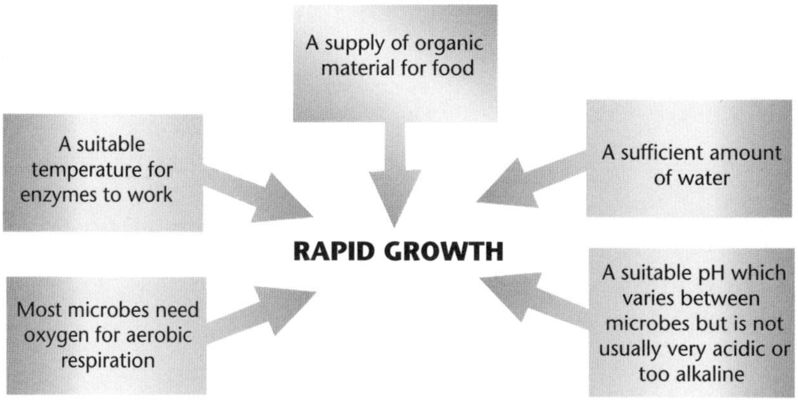

A supply of organic material for food

A suitable temperature for enzymes to work

A sufficient amount of water

RAPID GROWTH

Most microbes need oxygen for aerobic respiration

A suitable pH which varies between microbes but is not usually very acidic or too alkaline

Fig. 8.1

In ideal conditions microbes can reproduce very rapidly. This makes microbes very useful in the production of food and therefore a powerful tool in man's attempt to feed a rapidly increasing world population. Growth of microbes in stored food, however, results in large amounts of wastage.

Growing microbes

Microbes can be grown on a jelly-like material called agar in special containers called petri dishes. The agar contains dissolved nutrients to help the microbes grow.

> **KEY POINT** In order to grow a particular microbe, it is vital to use certain sterile techniques to prevent the entry of other microbes.

These precautions also help to prevent harmful bacteria from growing on the dish. The plates are also usually kept at 25°C which is below the preferred temperature of these bacteria.

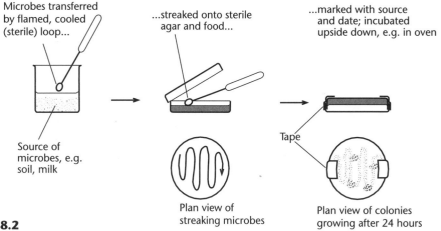

Fig. 8.2

This is the usual process of inoculating a petri dish with microbes:

Contamination of the petri dish by other microbes is prevented by:

- sterilising the loop in a flame before use
- sterilising the petri dish and agar before use
- only lifting one side of the lid to insert the loop
- sealing the lid with tape.

Louis Pasteur's experiments

Louis Pasteur was a famous French scientist. He worked with microbes and proved that it was bacteria that caused milk to turn sour. Up until then, people thought that living organisms, such as bacteria, could just appear in non-living material. Pasteur proved this to be false. By using a special-shaped flask, he showed that the bacteria that make broth go 'bad' come from the air.

This appearance of living organisms was called spontaneous generation.

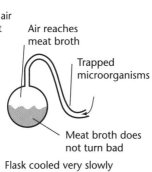

Fig. 8.3

1. Why do microbes need a suitable temperature in order to reproduce?
2. Why do human pathogens grow better at temperatures higher than 25°C?
3. Why is the wire loop cooled before dipping it into the source of microbes?
4. When Pasteur broke the long neck off the flask he found that the broth went 'bad'. Why was this?

1. If the temperature is too low, enzymes are inactivated, if it is too high enzymes are denatured;   2. Human body temperature is 37°C;   3. If the wire was too hot, it would kill the microbes;   4. The microbes were not caught in the neck and entered the flask in the air.

Preserving food

OCR A ^A
NICCEA
WJEC

Spoilage

Louis Pasteur showed that it is microbes that make food go bad or spoil. Food is dead organic material and so microbes, such as bacteria and fungi, will use it as a food source. They release enzymes on to the food. This changes the texture of the food and waste products may change the taste.

> **KEY POINT**
> The microbes that cause food to spoil are **decomposers**.

Other changes may also occur in the food that are not caused by microbes. An example of this is the oxidation of some of the substances in the food.

> Most types of food spoilage are harmless but make the food taste unpleasant. Some microbes, however, may cause food poisoning.

Methods of food preservation

To stop microbes spoiling food there are two main approaches:

- stop decay by preventing the microbes from reproducing; or

- stop decay by killing the microbes

Fig. 8.4

Preventing reproduction

> **KEY POINT**
> Many methods of food preservation work by removing one of the factors needed for microbes to reproduce.

Method of preservation	Details of the method	Factor targeted	Example of suitable food
freezing	food is rapidly cooled to −18°C	low temperatures inhibit enzyme action	most food, unless it has a high water content
drying	food is dried by hot air or by freeze drying	low water content stops microbes reproducing	fruit and vegetables, milk and coffee
jam making	food is mixed with sugar	high water potential of the jam draws water out of microbes stopping their reproduction	fruit

> Salting has the same effect as adding sugar.

Killing the microbes

> **KEY POINT** Many other methods of food preservation work by killing the microbes in the food.

Method of preservation	Details of the method	Action on microbes	Example of suitable food
canning	the food is heated to 121°C and sealed in cans	this temperature kills most microbes and spores	most foods
ultra heat treatment UHT	heated to 132°C for a short period of time by steam	high temperatures kill most microbes and spores. Short heating time means the taste is not changed too much	milk, fruit juices
irradiation	food is exposed to gamma radiation	high doses kill all microbes, low doses only kill some, does slow down ripening and sprouting	vegetables, fruit
chemical preservatives	addition of chemicals, such as sulphur dioxide and nitrates	chemical preservatives kill microbes	drinks and meat

Possible side effects

Chemical preservatives

Certain chemical preservatives have been shown to have certain side effects when eaten. Benzoic acid may cause hyperactivity in some children and scientists are worried that nitrates may increase the risk of cancer.

> It is important to weigh up these possible risks with the risks of food poisoning if they are not used.

Irradiation

The idea of exposing food to radiation has also worried some people. It does not make the food radioactive in any way but there is some concern that it may leave reactive free radicals in the food. The high doses needed to kill all the microbes cause changes in the taste and colour. Irradiation is therefore usually used in lower doses to kill some microbes.

PROGRESS CHECK

1. Name the two main groups of microbes responsible for food spoilage.
2. Why might food deteriorate even if there are no microbes present?
3. Freezing food slowly causes large ice crystals to form inside cells which breaks them open. What would happen to this food when it defrosts?
4. Why is food that has received UHT or been heated in cans, sealed after treatment?

1. Bacteria and fungi; 2. Oxidation by reaction with the oxygen in the air; 3. The food would become soft and mushy; 4. To prevent contamination by other microbes.

Traditional uses of microbes

Baking

> Vitamin C is often added to the flour to speed up the fermentation.

> Baking at high temperatures evaporates the alcohol.

flour, yeast, sugar (as nutrient for the yeast) and water are mixed

traditional fermentation – **proving** – takes several hours during which the yeast cells produce CO_2 so the dough increases in volume (a small amount of alcohol is also produced)

dough is divided into small sections and left for 30 minutes, then baked for 30 minutes

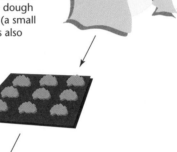

Fig. 8.5

Brewing

> The alcohol kills the yeast when it reaches about 14% by volume. To make spirits the alcohol must be distilled.

Barley is the source of sugar.

Barley is ground to separate husks from starchy interior.

Enzymes in the barley convert starch
Starch ⟶ maltose (sugar)

Husks are collected, dried and sold to make cattle food.

Roasted hops

Hops provide the flavouring

Hops are roasted

Yeast and extra sugar – different yeast strains for different beers, e.g. *Saccharomyces carlsbergensis* for lager

Fermentattion vat
• copper for beer (wood for wine)
• yeast + hops + sugar = wort
• here sugar is fermented to alcohol. The carbon dioxide escapes to form a froth which keeps the mixture anaerobic.

Typical fermentation temperature: 26 - 30°C

Fig. 8.6

Yoghurt and cheese

The lactic acid made by the bacteria causes the milk proteins to coagulate and thicken the yoghurt.

> **KEY POINT** The production of yoghurt and cheese are traditional ways of preserving milk.

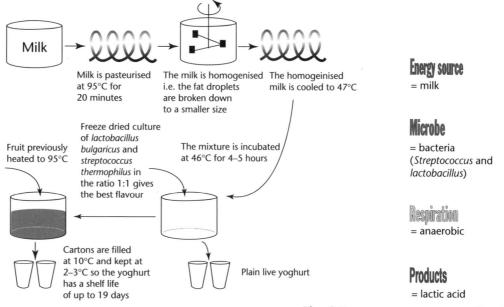

Milk is pasteurised at 95°C for 20 minutes

The milk is homogenised i.e. the fat droplets are broken down to a smaller size

The homogeinised milk is cooled to 47°C

Freeze dried culture of *lactobacillus bulgaricus* and *streptococcus thermophilus* in the ratio 1:1 gives the best flavour

Fruit previously heated to 95°C

The mixture is incubated at 46°C for 4–5 hours

Cartons are filled at 10°C and kept at 2–3°C so the yoghurt has a shelf life of up to 19 days

Plain live yoghurt

Fig. 8.7

Energy source
= milk

Microbe
= bacteria
(*Streptococcus* and *lactobacillus*)

Respiration
= anaerobic

Products
= lactic acid

Fig. 8.8

The more whey that is removed from the curd then the harder the cheese that is made.

The production of cheese is similar to the production of yoghurt. Bacteria are added to milk to produce lactic acid. This creates the correct pH for the enzyme rennin to coagulate the milk proteins. This solid material is called the curd and is separated from the liquid whey.

Vinegar

Energy source
= alcohol

Microbe
= bacteria
(*Acetobacter*)

Respiration
= aerobic

Products
= ethanoic acid

Fig. 8.9

Vinegar is made by trickling beer or cider through wood shavings, which are coated with the bacteria. The vinegar can then be removed from the bottom of the container.

Soy sauce

Energy source
= soya beans

Microbe
= fungi
(*Aspergillus*)
then yeast and bacteria
(*Lactobacillus*)

Respiration
= anaerobic

Products
= soya sauce

Fig. 8.10

1. Why does heating the bread in an oven cause the bread to rise further?
2. Why can riper grapes make wine with more alcohol?
3. The loosely packed wood shavings let air into the vinegar-making container. Why is this important?

PROGRESS
CHECK

1. Why does heating the bread in an oven cause the bread to rise further?
2. Why can riper grapes make wine with more alcohol?
3. The loosely packed wood shavings let air into the vinegar-making container. Why is this important?

1. The trapped CO_2 expands when heated; 2. Riper grapes contain more sugar; 3. Provides oxygen as the process is aerobic.

Modern uses of microbes

AQA
Edexcel A   Edexcel B
OCR A ^A   OCR A ^B
NICCEA
WJEC

Fermenters

Microbes can now be grown in large amounts very rapidly. In order to do this they are grown in large vessels called **fermenters**.

KEY
POINT

Fermenters are designed to grow large quantities of a single type of microbe in carefully controlled conditions.

Although they are called fermenters, the microbe may be growing using oxygen, i.e. aerobically. This is not fermentation.

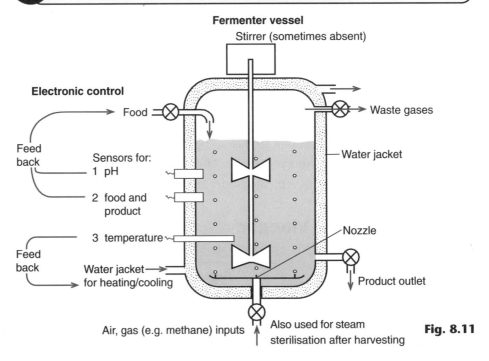

Fig. 8.11

A pure culture of the microbe grows rapidly because:

Sterilisation

Kills any unwanted microbes that might compete for the food and produce unwanted products.

Controlled condition

The temperature, food levels and oxygen levels are all monitored and kept at ideal levels. Aerobic reactions will need oxygen, anaerobic will not.

SCP and mycoprotein

Microbes, such as mushrooms, have been used as food for centuries but it was not until the last thirty years that their real potential has been investigated. Scientists began to realise that microbes have a number of advantages over meat as a source of protein:

- they have a faster growth rate than animals

- the microbes can be grown on waste products, such as whey

- they have a higher fibre content

- they have a lower fat content

- they can provide protein without killing animals.

> **Whey is the waste from cheese production.**

Two main types of microbes have been grown as protein sources. These are a fungus called *Fusarium* and various bacteria. The bacteria are mainly used for cattle feed and the fungi for human consumption.

> **KEY POINT**
>
> The protein produced by these microbes is called **Single Cell Protein (SCP)** and if it is produced specifically by a fungus it is often called **mycoprotein.**

> **The substrate can be any carbohydrate source, such as the waste from cheese making.**

> **The food processing makes the food look and taste more appealing.**

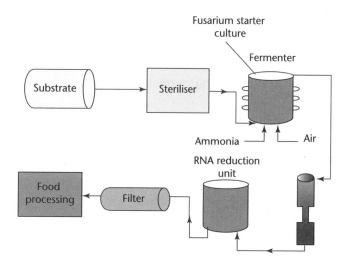

Fig. 8.12

There are some disadvantages of SCP:

- it has less variety of amino acids than many types of meat

- it has a high level of RNA that can be harmful to humans, this has to be reduced

- the taste is rather bland without flavourings.

PROGRESS CHECK

1. What is the purpose of the stirrer in the fermenter?
2. Why did mycoprotein have to undergo years of testing before humans were allowed to eat it?
3. To simulate the texture of meat, SCP made from fungi is used rather than bacteria. Why is this?

1. The stirrer keeps the microbes mixed with the food and oxygen and distributes the heat; 2. To make sure that it was safe to eat; 3. The hyphae in the fungus give the mycoprotein a texture like meat.

Sample GCSE question

1. The following graph shows some of the changes that occur during the production of wine.

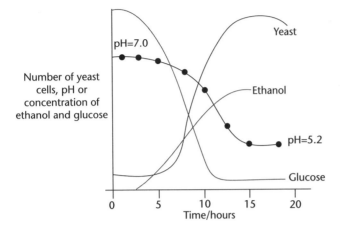

(a) Write a word equation for the reaction taking place in the mixture. **[2]**

glucose ✓ → carbon dioxide + ethanol ✓

> This is the equation for anaerobic respiration in yeast or plants.

(b) Give **two** reasons why the number of yeast cells levels off after about 13 hours. **[2]**

1. The yeast starts to run out of glucose for food ✓.
2. The yeast is killed by the build up of ethanol ✓.

> The decrease in pH may also affect the yeast.

(c) The temperature of the mixture was kept at about 25°C, this was above the temperature of the room.

(i) Explain why the mixture was kept above room temperature. **[1]**

To increase the rate of reaction ✓.

(ii) At the start of the reaction, this involved heating the mixture but then later on, the mixture needed cooling to keep it at 25°C.

Explain why it needed cooling later in the reaction. **[1]**

The fermentation of the yeast produced heat. This increased the temperature of the mixture to above 25°C ✓.

(iii) Why might the reaction slow down if the temperature became too high? **[2]**

The reaction is controlled by enzymes in the yeast cells ✓. High temperatures may change the shape or denature the enzymes, so the reaction would slow down ✓.

> Remember not to say that 'enzymes are killed'.

Exam practice questions

1. Look at the diagram showing detail of a process for making Single Cell Protein.

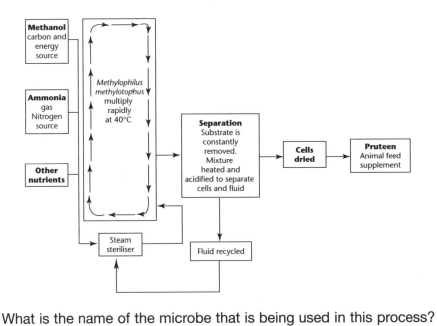

(a) What is the name of the microbe that is being used in this process? **[1]**

(b) Why does the microbe need a nitrogen source? **[2]**

(c) One of the other nutrients is oxygen. What is this needed for? **[1]**

(d) Explain one way in which contamination by other microbes is prevented. **[1]**

(e) What are the dangers of contamination by other microbes? **[2]**

(f) Name a microbe that may have been used if the protein had been intended for human consumption. **[1]**

2. The following diagram shows the main steps in yoghurt making.

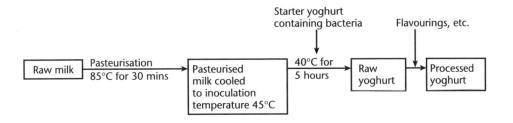

(a) Why is the raw milk pasteurised before use? **[1]**

(b) Why is the starter cooled to 40°C before the bacteria are added? **[1]**

(c) What are the bacteria feeding on in the milk and what is their main product? **[2]**

Microbes, waste and fuel

The following topics are covered in this section:

● *Sewage treatment*   ● *Fuels*
● *Other industrial uses of microbes*

LEARNING SUMMARY

After studying this section you should be able to:

● describe how microbes are used to treat sewage
● explain how microbes can be used to generate new types of fuel.

KEY POINT

The ability of microbes to breakdown dead organic material has been used for some time in the treatment of sewage. The waste products of this decomposition were once considered to be dangerous by-products but now are used to produce fuels.

Sewage treatment

Edexcel A   Edexcel B

WJEC

Sewage is made up of a number of different components but includes toilet flushings, kitchen wastes and water from washing. This is mixed with some industrial waste, which by law has to receive some treatment before it is allowed to enter the sewers.

KEY POINT

Sewage is treated to make it safe by using microbes to break down the organic substances.

The spread of diseases is covered in Chapter 10 and Eutrophication in Chapter 13.

Proper sewage treatment is important for two main reasons:

● it removes pathogens and so stops them infecting other people

● it allows the water to be released into rivers without the dangers of eutrophication.

There are two main types of sewage treatment:

● **Biological filter** (trickling filter): this involves trickling the sewage through a bed of stones that are coated with a community of microbes. The microbes break-down and feed on the organic matter.

● **Activated sludge**: this can deal with larger amounts of sewage and involves adding microbes to the sewage. It is then held in large tanks whilst air is bubbled through.

Both processes rely on the sewage being allowed to settle first so that the larger particles are removed as sludge.

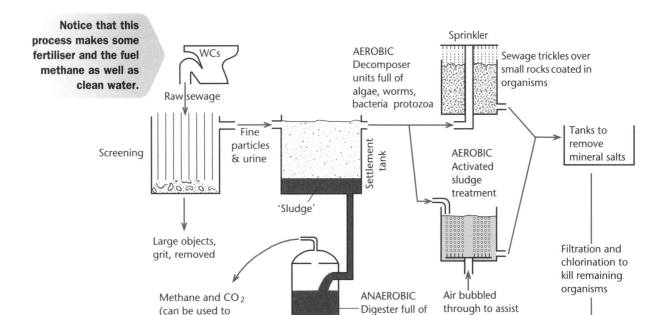

Notice that this process makes some fertiliser and the fuel methane as well as clean water.

Fig. 9.1

Fuels

AQA
WJEC

Fossil fuels have been used for centuries but the increasing demand for energy has led to fossil fuel supplies running short and widespread pollution. Recently, new fuels have been developed.

> **KEY POINT** Microbes are used to convert a range of plant products and waste materials into fuels.

In India and China many families have a digester making biogas for fuel.

Biogas is mainly methane and is made by a collection of different bacteria. They ferment waste material such as animal dung or the sludge from sewage plants as mentioned above.

The waste is fermented in a covered tank called a **digester**.

In Brazil there are over one million cars modified to use gasohol.

Gasohol is a mixture of 80% petrol and 20% ethanol. The ethanol is produced by fermenting sugar from sugar cane or starch. When mixed with petrol it can be used in cars and has advantages over using pure petrol:

● it produces less pollutants, such as sulphur dioxide and the carbon dioxide released by the ethanol is balanced by the uptake into sugar cane

● the dry pulp left over from the sugar cane can be used as an animal feed.

Other industrial uses of microbes

OCR A B
WJEC

Bacteria and fungi are also grown in fermenters to produce large amounts of enzymes. The enzymes are then used for various processes:

- washing powders – enzymes such as proteases and lipases are produced by bacteria and used in biological washing powders to digest stains

- extracting fruit juices – enzymes such as pectinases are made by fungi and are used to digest plant tissue to speed up the extraction of fruit juices.

Nowadays, the microbes may be genetically engineered in order to produce certain enzymes: see Chapter 11.

PROGRESS
CHECK

1. Why is air bubbled through the sewage in the activated sludge method of treatment?
2. Why does burning methane or ethanol cause less acid rain than the burning of fossil fuels?
3. It is recommended that biological washing powder is used at fairly low temperatures. Why is this?
4. How can washing clothes at low temperatures help to conserve the environment?

1. To provide the microbes with oxygen for aerobic respiration; 2. Methane or ethanol do not contain impurities of sulphur and so do not give rise to sulphur dioxide when burnt; 3. High temperatures would lead to the enzymes in the powder denaturing; 4. Washing at low temperatures requires less energy and so requires less fossil fuel to be burnt.

Sample GCSE question

1. The diagram shows a typical digester that is used in India to produce biogas.

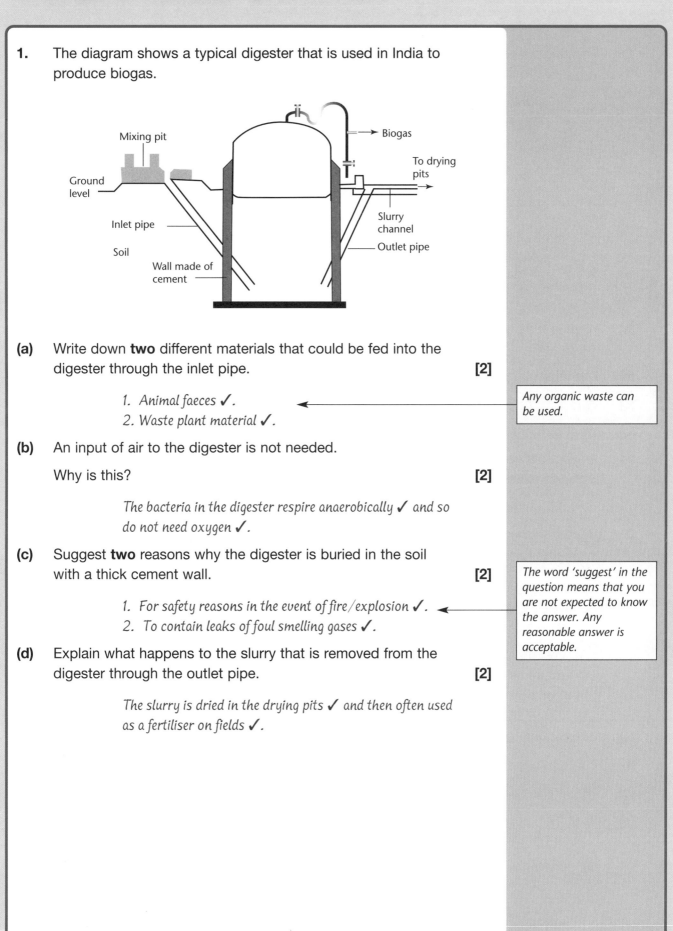

(a) Write down **two** different materials that could be fed into the digester through the inlet pipe. **[2]**

> 1. Animal faeces ✓.
> 2. Waste plant material ✓.

Any organic waste can be used.

(b) An input of air to the digester is not needed.

Why is this? **[2]**

> The bacteria in the digester respire anaerobically ✓ and so do not need oxygen ✓.

(c) Suggest **two** reasons why the digester is buried in the soil with a thick cement wall. **[2]**

> 1. For safety reasons in the event of fire/explosion ✓.
> 2. To contain leaks of foul smelling gases ✓.

The word 'suggest' in the question means that you are not expected to know the answer. Any reasonable answer is acceptable.

(d) Explain what happens to the slurry that is removed from the digester through the outlet pipe. **[2]**

> The slurry is dried in the drying pits ✓ and then often used as a fertiliser on fields ✓.

Exam practice questions

1. The diagram shows an outline of the processes involved in the treatment of sewage.

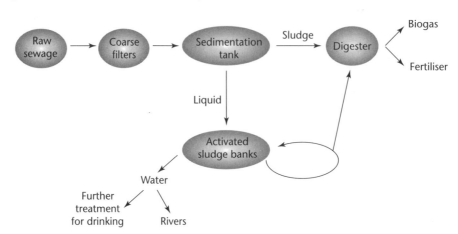

(a) What is the purpose of passing the sewage through coarse filters at the start of the treatment? **[1]**

(b) Describe what happens in the activated sludge tank. **[3]**

(c) Why is some of the sludge recycled back into the treatment tanks? **[1]**

(d) What further treatment is needed before the water can be used for drinking? **[1]**

(e) What could be the possible consequences if sewage was released into rivers without treatment? **[3]**

2. Read the following passage about gasohol and answer the questions that follow.

> In 1972 the price of oil went up dramatically and this made many countries look at ways of producing alternative fuels. A programme was set up in Brazil to produce ethanol by fermentation of sugar cane, followed by distillation. The ethanol can be used on its own or in a mixture called gasohol.
>
> At present, ethanol is more expensive than petrol but this may not be the case in the future. At present Brazilian biologists are working on genetically modified yeast that will digest and ferment starch.

(a) What is ethanol mixed with in order to produce gasohol? **[1]**

(b) Apart from the cost, write down one advantage of using gasohol rather than petrol. **[1]**

(c) Why is the cost of petrol likely to increase in the distant future? **[1]**

(d) **(i)** Why is producing gasohol more likely to be feasible in hot countries, such as Brazil, rather than in Britain? **[1]**

(ii) Why could the production of the genetically modified yeast change this? **[2]**

Microbes and disease

The following topics are covered in this section:

- **Causes of disease**
- **Antiseptics, antibiotics and painkillers**
- **Vaccines and their production**
- **Spread of disease**
- **Use of antibiotics**

LEARNING SUMMARY

After studying this section you should be able to:

- **understand the causes of disease and how they are spread**
- **understand about various types of disease and how they are spread**
- **understand the use of antiseptics and antibiotics**
- **understand how vaccines work**
- **understand about some plant diseases and methods of control.**

KEY POINT

Microbes are small microscopic organisms. Most microbes are perfectly harmless and some microbes are very useful. We use them in all sorts of ways, including making food such as bread and for making antibiotics to attack disease causing, or pathogenic microbes. It is only a few microbes that belong to this group of pathogenic organisms.

Causes of disease

AQA
Edexcel A   Edexcel B
OCR A ᴬ   OCR A ᴮ
NICCEA
WJEC

Diseases can either be *caught from other people, inherited from our parents, caused by our environment,* or even *caused by parasites.*

Microbes that cause disease are called pathogens.

Remember viruses are much smaller than bacteria.

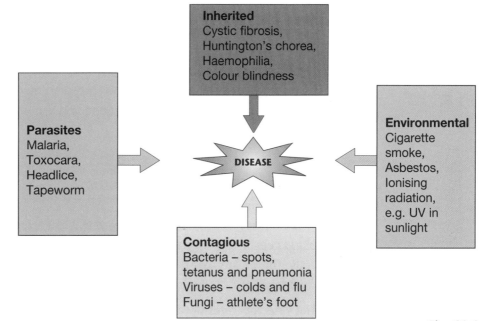

Inherited
Cystic fibrosis, Huntington's chorea, Haemophilia, Colour blindness

Parasites
Malaria, Toxocara, Headlice, Tapeworm

DISEASE

Environmental
Cigarette smoke, Asbestos, Ionising radiation, e.g. UV in sunlight

Contagious
Bacteria – spots, tetanus and pneumonia
Viruses – colds and flu
Fungi – athlete's foot

Fig. 10.1

PROGRESS CHECK

1. List three contagious diseases and for each one say how it is transmitted.
2. List three different types of non-infectious disease, and for each one state how it is acquired.

2. Examples should include a genetic example and an environmental example.
1. Tetanus or pneumonia – bacteria;   Colds or flu – viruses; Athletes foot – fungi;

Spread of disease

Famous scientist proves bacteria theory

Louis Pasteur has finally proven that food goes bad because of small microbes called bacteria. He boiled some gravy in a swan necked flask and the gravy stayed fresh for weeks. He believed that bacteria from the air, fell into and were trapped in the swan neck. Sure enough, when he tilted the flask so that the gravy went into the neck of the flask, the gravy went bad in just a couple of days.

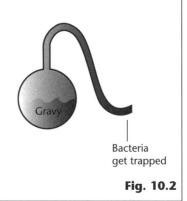

Gravy

Bacteria get trapped

Fig. 10.2

Pasteur and disease

In the early 1800s, people thought that 'life' could appear out of thin air and that people became ill because of 'bad air'. Louis Pasteur was able to show that food went 'bad' because of bacterial decay. This was a revolutionary idea at the time. Pasteur was able to show that:

● fermentation was caused by yeast and certain bacteria could ruin the process

● sterilised food stayed fresh, until bacteria were allowed to get to it.

Pasteur and Robert Koch were later able to show that diseases such as anthrax were caused by similar microorganisms.

HIV

Most diseases spread because they are infectious. The **Human Immunodeficiency Virus (HIV)** is spreading quickly through some populations. The virus is easily damaged and can only be passed on by direct body fluid to body fluid contact.

> **KEY POINT** HIV positive means a person is infected with the HIV virus. It does not mean that they have AIDS.

When a person becomes infected with HIV, the virus RNA enters the white blood cells. These are the same cells that are trying to find and destroy the virus. This means that the virus is in the very place that the white blood cells cannot get to.

This long time period allows the virus to spread to many other people before the host becomes ill and dies.

The virus replicates slowly by instructing the cell to make copies of the virus. The cell dies and releases hundreds of copies of the virus to infect other cells. Initially, although the person could now be spreading AIDS, it does not cause any problems for the infected person. Eventually however, so many white cells are destroyed that the person can no longer fight off infections.

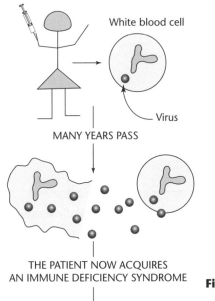

White blood cell

Virus

MANY YEARS PASS

THE PATIENT NOW ACQUIRES
AN IMMUNE DEFICIENCY SYNDROME

Fig. 10.3

As white blood cells are destroyed, our immune system becomes weaker and can no longer protect us from other diseases. This collection of diseases is called a **syndrome**. It is not the HIV infection that directly leads to death, but the syndrome of diseases that are caused as a result of the HIV infection.

AIDS

Fig. 10.4

The spread of HIV can be prevented by:

Safe sex
use a condom

Not re-using
hypodermic needles

Having fewer
sexual partners

Fig. 10.5

Parasites

Parasites are *larger* than microbes and can usually be seen with the *naked eye*.

Toxocara is a worm that is small and white. It is sometimes called a 'round worm'. It usually lives in the gut of cats and dogs.

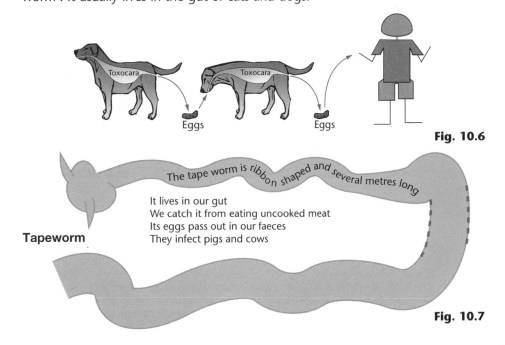

Toxocara

Toxocara

Eggs

Eggs

Fig. 10.6

The tape worm is *ribbon shaped and several metres long*

It lives in our gut
We catch it from eating uncooked meat
Its eggs pass out in our faeces
They infect pigs and cows

Tapeworm

Fig. 10.7

Fortunately, tapeworm is rare in Great Britain because we have a good system for inspecting meat, and sewage is treated and not used on fields as a fertiliser. This breaks the chain in the life-cycle of the tapeworm, so it is not passed back to humans.

Tapeworms can also be caught from many other animals, e.g. fish.

Its spread can be prevented by:

● ensuring that raw meat is well cooked

● ensuring that all meat is carefully inspected

● ensuring good sanitation and hygiene and not using human sewage as a fertiliser.

Malaria is a disease caused by a small parasite that lives in the blood. It is spread by the blood-sucking anopheles mosquito.

> **KEY POINT**
>
> An organism that transmits a disease is called a **vector**.

It can be controlled by:

● **drugs** – holiday makers should take drugs before they go on holiday so that if they are bitten by the mosquito, the drug is in the blood ready to kill the parasite.

● **chemical insecticides** – these kill the mosquito so the disease cannot be spread

● **draining swamps** – this removes the breeding grounds for the mosquito

● **stocking lakes with fish** – the fish eat the larvae of the mosquito

● **using mosquito nets at night** – this prevents the mosquito biting and passing on the parasite.

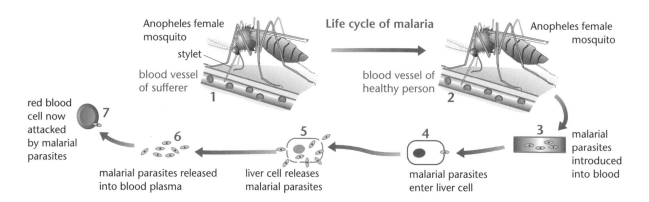

Fig. 10.8

---

1. State the name of the person who proved that disease is caused by microbes.
2. Explain how he managed to prove that microbes caused food to go bad.
3. Explain how HIV is transmitted and how its spread can be prevented.
4. State one parasite and explain its life cycle.

PROGRESS CHECK

1. Louis Pasteur; 2. He used a swan necked flask to show that only when bacteria reached food, would the food go bad; 3. Direct tissue fluid to tissue fluid contact, e.g. unprotected sex and dirty hypodermic needles. Use condoms and clean needles; 4. E.g. tapeworm – contaminated meat, tapeworm grows in gut, releases eggs in faeces, cows eat eggs, meat becomes contaminated.

Antiseptics, antibiotics and painkillers

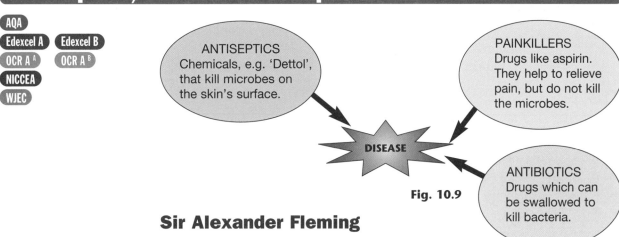

ANTISEPTICS
Chemicals, e.g. 'Dettol',
that kill microbes on
the skin's surface.

PAINKILLERS
Drugs like aspirin.
They help to relieve
pain, but do not kill
the microbes.

DISEASE

ANTIBIOTICS
Drugs which can
be swallowed to
kill bacteria.

Fig. 10.9

Sir Alexander Fleming

Fleming discovered the antibiotic **penicillin**.

He noticed that a culture plate of bacteria were being killed off by an unknown substance. One of his plates was contaminated by a mould called penicillium and the mould was killing off the bacteria on the plate. Most people would have simply thrown the plate away, but Fleming realised that this was a remarkable discovery.

> Fleming could see that where the mould was growing, it was killing the bacteria around it. He concluded that the mould must be producing a chemical that was diffusing outwards and killing the microbes.

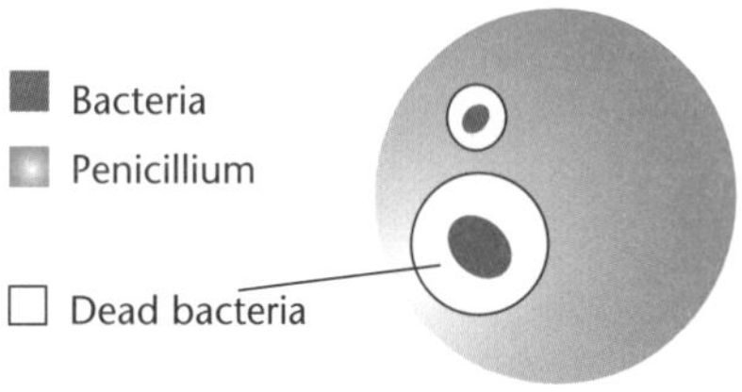

■ Bacteria
▨ Penicillium
□ Dead bacteria

Fig. 10.10

> The work of Fleming, Florey and Chain is a good example of **Science and Evidence.**

Fleming worked closely, with two other scientists, called **Florey** and **Chain**. They isolated the penicillin from the mould, and were awarded the Nobel Prize for Medicine.

Penicillin was regarded as a miracle drug and saved the lives of thousands of wounded soldiers during World War II.

The normal scientific method of hypothesis, experimentation and observation did not happen in this discovery. Instead a chance observation by Fleming, which could easily have been overlooked, opened up a whole new branch of scientific discovery. There must be many examples where new discoveries are missed because a scientist fails to observe an apparently insignificant and unrelated fact.

PROGRESS CHECK

1. Describe the difference between antiseptics, antibiotics and painkillers.
2. Describe the role played by Fleming, Florey and Chain, in the discovery of penicillin and how the scientific method was replaced by good fortune and a prepared mind.

1. Painkillers remove painful symptoms, but do not kill microbes or cure the disease. Antiseptics are used externally to prevent infections in wounds. Antibiotics are drugs taken internally to kill bacteria that cause infection;   2. The answer should explain how Fleming discovered penicillin and also how good fortune replaced the normal scientific method.

Use of antibiotics

AQA
Edexcel A   Edexcel B
OCR A ^A   OCR A ^B
NICCEA
WJEC

There are now many different kinds of **antibiotics**. Most are made from different kinds of fungi and are used to kill a wide range of different kinds of bacteria.

> **KEY POINT**   Antibiotics are only effective against bacteria. They do not work against viruses.

Antibiotics must be capable of killing bacteria without harming body tissue.

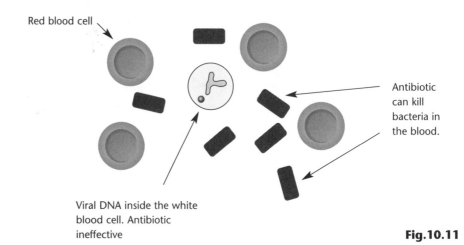

Red blood cell

Antibiotic can kill bacteria in the blood.

Viral DNA inside the white blood cell. Antibiotic ineffective

Fig.10.11

It is very difficult to develop a drug that will destroy a virus, but not also damage body tissue.

The problem of antibiotic resistance

The summer holidays last for about 42 days. Work out how much you would earn if you had a job that paid 1p on the first day, and doubled your earnings each day, for 42 days.

Unlike most other organisms, bacteria multiply very rapidly. A single bacterium can divide every 20–30 minutes. This means that in only a few hours, one bacterium can give rise to many thousand. This enables a bacterium to infect a person and within hours, have reproduced to produce enough bacteria to cause symptoms of the disease.

This rapid rate of asexual reproduction allows bacteria to quickly develop antibiotic resistance.

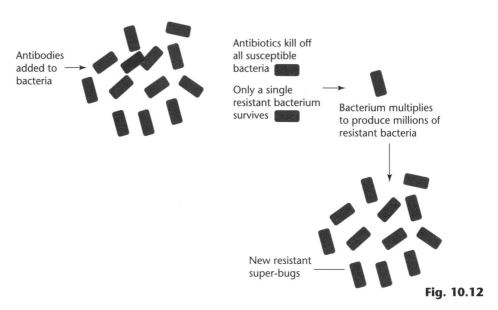

Antibodies added to bacteria

Antibiotics kill off all susceptible bacteria

Only a single resistant bacterium survives

Bacterium multiplies to produce millions of resistant bacteria

New resistant super-bugs

Fig. 10.12

Antibiotic resistance is now a major problem. Dangerous bacteria are becoming resistant to more and more antibiotics. It is now so serious that one bacteria, called MRSA, is resistant to all but one known antibiotic and is virtually incurable. Scientists are racing against time to find new antibiotics against these new 'super-bugs'.

Doctors try to prevent antibiotic resistance by:

● reducing the use of antibiotics. Patients often want the doctor to give them antibiotics for viral diseases against which antibiotics are useless.

● not using antibiotics in animal feed. In the past they have been used as a growth promoter.

● using combinations of antibiotics. This ensures that if a single bacterium survives because it is resistant to the antibiotic, another different antibiotic kills it off before it can start to reproduce.

● ensuring that the patient completes the course of treatment. Patients often feel much better after a few days and fail to finish the tablets. This leaves the more resistant bacteria to survive and multiply.

Phage technology could provide an answer to antibiotic resistance. Bacteriophages are viruses that attack bacteria. By culturing phages that attack pathogenic bacteria, and evolve just as quickly as the bacteria do, scientists hope to stay one jump ahead in the 'germ warfare' race.

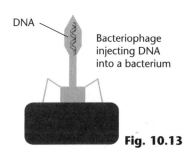

DNA

Bacteriophage injecting DNA into a bacterium

Fig. 10.13

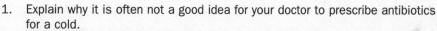

1. Explain why it is often not a good idea for your doctor to prescribe antibiotics for a cold.
2. Explain why bacteria can evolve to be resistant to antibiotics in only a few years, while evolutionary change in humans takes millions of years.
3. Explain the process to show how a species of bacteria can become resistant to a new antibiotic.
4. MRSA is a super-bug that is resistant to almost all antibiotics. Explain how we can prevent other bacteria from becoming resistant to antibiotics.

PROGRESS CHECK

1. Colds are often caused by viruses. Because these exist inside cells, antibiotics are incapable of getting to them to destroy them.   2. Humans reproduce about once every twenty years or so. Bacteria reproduce one every twenty minutes. This allows them to evolve much more quickly;   3. One bacterium in every few million may have a mutation that gives them immunity to a new antibiotic. The antibiotic will kill all the susceptible bacteria, leaving only the resistant one alive. This will then breed to produce millions of resistant bacteria;   4. Use less antibiotics. Use multiple antibiotic treatments. Finish a course of treatment.

Vaccines and their production

Vaccines provide protection against disease. They are injected into the body and cause the body to produce antibodies to combat the real disease when the body comes into contact with it.

The first person to produce a vaccine was Edward Jenner. He developed a vaccine against **smallpox**.

Edward Jenner was a doctor who lived in the late 1700s. He was often called upon to treat milk-maids who had caught a disease called cowpox. It was a mild disease that caused blisters to appear on the milk-maids' hands. Jenner noticed that the milk-maids who had caught cowpox, never caught the much more serious disease of smallpox.

> A good example of Science and Evidence.

He decided to do an experiment. He persuaded a mother to let him inject her son with cowpox. The boy caught the disease and then recovered. Jenner then injected the boy with deadly smallpox. Fortunately for Jenner and the boy, he did not contract smallpox. Jenner realised that by injecting people with the harmless cowpox, he could protect them from the much more deadly smallpox.

Louis Pasteur developed a vaccine against rabies. He injected the rabies virus into a rabbit's brain. He then removed the brain tissue, dried it, and injected it into another rabbit. He repeated this several times until the virus had

> Another example of Science and Evidence.

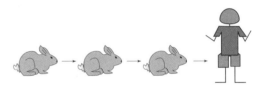

Fig. 10.14

become weakened and no longer caused disease in the rabbit. He then injected some of this weakened virus into a person as a vaccine. Every few weeks he injected the same person with a stronger and stronger form of the virus, until he was able to inject them with the virulent pathogen, with no ill effect.

Different forms of vaccine

- Jenner's vaccine worked because the cowpox virus is so similar to the smallpox virus that our bodies cannot tell the difference.

- Pasteur's vaccine worked because he weakened the virus so much that it was no longer capable of invading our cells and causing the disease.

- Other vaccines can be made from dead bacteria or even parts of the bacteria's outer coat.

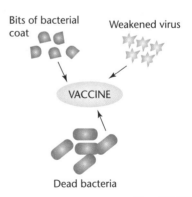

Fig. 10.15

How vaccines work

When we catch a disease, the microbe **(antigen)**, enters our body and starts to multiply. This causes our bodies to make chemicals to destroy it **(antibodies)**. It is a race between making antibodies quickly enough to destroy the microbe, and being destroyed by the microbe.

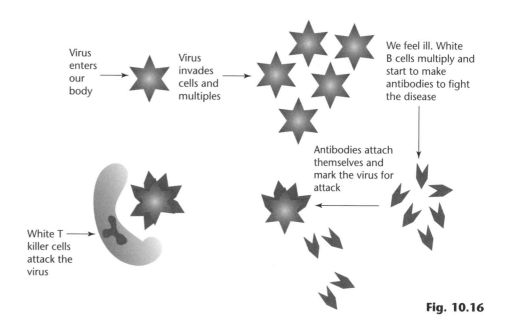

Virus enters our body → Virus invades cells and multiples → We feel ill. White B cells multiply and start to make antibodies to fight the disease

Antibodies attach themselves and mark the virus for attack

White T killer cells attack the virus

Fig. 10.16

Memory cells remain in the blood ready to produce more antibodies if the antigen enters the body a second time.

 Remember the word pathogen means disease causing and virulent means a dangerous disease.

The disease can be prevented by **vaccination**. Dead or inactivated microbes, are injected into our bodies. We then make antibodies to fit the shape of the harmless microbe. Later when the real virulent pathogen enters our body, we have the memory cells and antibodies ready to destroy it before it can multiply and cause disease.

> **KEY POINT** When we make our own antibodies to fight a disease, it is called active immunity.

This is another good example of Ideas and Evidence.

An example is the **MMR** vaccine. It protects children against measles, mumps and rubella.

Some people think that the vaccine may be responsible for causing brain damage in the young children who are vaccinated. Parents need to make an informed choice about whether to have their children vaccinated. They need to weigh the proven risks of catching measles, mumps and rubella, which may result in life-long disability and even death, against the unproven risks of having an adverse reaction to the vaccine.

Passive immunity

Vaccination to prevent disease is always the best option. Sometimes, however, vaccination is too late. Once we contract a disease vaccinations are not much use. By the time the vaccine has caused us to make antibodies we would either be dead, or have already made our own antibodies against the real disease. When this happens, the only solution is to be injected with ready-made antibodies that have been produced by and collected from someone else.

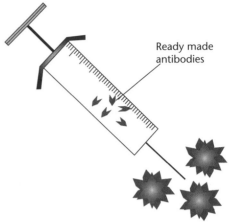

Ready made antibodies

Fig. 10.17

Advantages

- Injection with ready-made antibodies gives instant protection against the disease as the body does not have to take time to make them.

Disadvantages

- The protection does not last very long. Once the antibodies have gone we are once more susceptible to the disease. The only solution is active immunity where we learn to make our own antibodies.

> If you have never had a tetanus jab and you injure yourself, your doctor may inject you with ready-made antibodies. You will still need a normal tetanus vaccine later to give long-term protection.

Why vaccines do not always give long-term protection

It seems strange that one vaccination will give us long-term protection to a disease like tetanus, and yet we cannot get long-term protection against diseases like influenza and the common cold. This is because the virus has the ability to mutate and change the shape of its outer coat. When the virus returns some time later, our antibodies no longer fit the shape of the virus and we have to start making antibodies all over again.

> This is why flu jabs only last for one year. Each year a new vaccine has to be made.

Virus mutates and changes shape. Antibody no longer fits

New antibody has to be made to fit the shape of the virus

Fig. 10.18

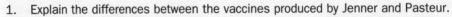

1. Explain the differences between the vaccines produced by Jenner and Pasteur.
2. Explain whether you think Jenner would be allowed to do his kind of research, using human guinea pigs, today.
3. Explain why we can catch a common cold, every year, but only catch measles once.
4. Explain the differences between active and passive immunity.
5. Explain the role played by antibodies in our immune system.

PROGRESS CHECK

1. Jenner's smallpox vaccine uses cowpox, a different but similar organism, as the vaccine. 2. No – with good ethical and moral reasons given; 3. Measles virus does not mutate, the cold virus does. This means that our antibodies no longer fit and new ones have to be made; 4. Passive immunity uses ready made antibodies – quick but short lasting. Active immunity is when we make our own antibodies – slow but last a long time; 5. Antibodies attach themselves to the antigen so that white blood cells can recognise it as foreign and destroy it.

Sample GCSE question

1. John fell and hurt his knee. His father put some antiseptic on the wound.

(a) Explain the job of the antiseptic. **[2]**

> *Antiseptics are used externally ✓ to kill microbes ✓.*

Remember – look to see how many marks a question is worth. You need to make at least that number of points in your answer.

(b) John does not take care of his wound. It becomes infected.

Explain the role played by John's immune system. **[3]**

> *Antibodies are made to fit the shape of the microbes ✓.*
> *White blood cells are now able to recognise the antigen ✓.*
> *White blood cells destroy the antigen ✓.*

Questions that start with 'explain' usually require at least two points in your answer.

(c) His mother decides to take him to the doctor. The doctor gives John some antibiotics and a tetanus jab. He says the jab will give instant protection but not last for long, He tells John that he must return in a few weeks for another vaccination.

(i) State with reasons, what type of microbe was likely to be causing the infection. **[2]**

> *Bacteria ✓ because the doctor gave him antibiotics which do not work against viruses ✓.*

(ii) Suggest what was likely to be in the injection that the doctor gave to John. **[2]**

> *Ready-made antibodies ✓ against tetanus ✓.*

(iii) Explain why John had to return a few weeks later for another jab. **[2]**

> *To get active immunity ✓ by making his own antibodies ✓.*

(d) John knows that the first vaccine was discovered by Edward Jenner. Jenner produced the vaccine for smallpox by experimenting on a young boy.

This question contains some Ideas and Evidence where you have to make a judgement about the morality of science in the past.

Explain how Jenner made his discovery and why he would not be allowed to do this today. **[5]**

> *Milk-maids who catch cowpox are immune to smallpox ✓.*
> *Jenner infected a young boy with cowpox who caught cowpox and recovered ✓.*
> *Jenner then infected the boy with smallpox ✓.*
> *The boy stayed healthy ✓.*
> *Some reference in the answer to the immoral way that the boy was used as a 'guinea pig' ✓.*

This is an extended answer and requires you to write several sentences in a logical order.

Exam practice questions

1. In the 1800s, people thought that bacteria appeared spontaneously out of thin air. Louis Pasteur proved that decay was caused by microorganisms called bacteria.

He used the following apparatus to prove his theory.

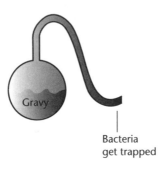

(a) Explain why the gravy stayed fresh when boiled in the flask **[1]**

(b) Explain the purpose of the two bends in the flask. **[2]**

After a few weeks, Pasteur shook the flask so that some of the gravy went into bend A. He then left the flask for another few days.

(c) Explain why the gravy now went bad. **[2]**

Pasteur went on to prove that some diseases were also caused by microorganisms.

(d) State two other diseases not caused by microorganisms and explain how they are caused. **[4]**

(e) AIDS is a new disease that did not exist when Pasteur was alive.

Explain why our immune system finds it difficult to destroy the virus that causes AIDS. **[2]**

2. Mary is going on holiday to India. She goes to her doctor for some vaccinations.

(a) Explain why Mary has her vaccinations some weeks before she goes away on holiday. **[1]**

(b) The doctor says she also needs to take anti-malarial tablets two days before her holiday and not wait until she gets there.

Explain why. **[2]**

(c) Mary returns from holiday with a bacterial infection. Her doctor is concerned about the bacteria developing antibiotic resistance.

Explain why he gives her a mixture of two antibiotics. **[2]**

(d) State two other ways that antibiotic resistance can be prevented. **[2]**

Chapter 11

Genetics and genetic engineering

The following topics are covered in this section:

- **Structure of DNA**
- **Protein synthesis**
- **Mutations**
- **Gene detection**
- **Genetic engineering**
- **Genetic fingerprints**
- **Cloning**

LEARNING SUMMARY

After studying this section you should be able to:

- understand the structure of DNA (deoxyribonucleic acid)
- understand how DNA makes proteins
- understand how the genes that cause genetic diseases can be found
- understand how genetic engineering works.

KEY POINT

The science of genetics is one of the most rapidly growing sciences. With the introduction of genetic engineering society will find itself having to make many difficult choices. It offers many advances in the elimination of starvation and disease but care is required in how it will be used.

Structure of DNA

AQA

Edexcel A   Edexcel B

OCR A ᴬ   OCR A ᴮ

NICCEA

WJEC

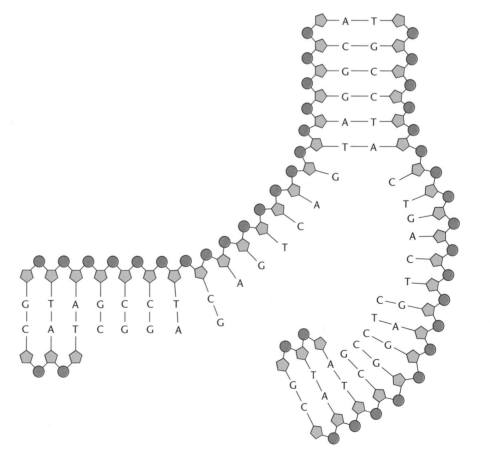

Fig. 11.1

DNA – key facts

- The language of DNA is universal and understood by all living organisms.
- DNA has two jobs, code for information and copy itself.
- The language has four letters. They are the bases **A**, **T**, **C** and **G**.
- Bases go together in pairs, **A** and **T** and **C** and **G**. This makes DNA a double strand.
- The sequence of bases codes for information. Just like the sequence of letters in English.
- Only one strand carries the message. This is called the 'sense' strand.
- The bases are held together by a ribose sugar ⬠ attached to a phosphate molecule ⬤ .
- DNA copies itself just before the cell divides. This ensures each new cell has an exact copy.
- To copy itself, DNA 'unzips' between the bases. Because **A** only pairs with **T** and **C** with **G**, a new copy is quickly made.

Protein synthesis

AQA
Edexcel A   Edexcel B
OCR A ᴬ   OCR A ᴮ
NICCEA
WJEC

DNA is much too precious to be allowed outside the cell. It is kept safe inside the nucleus. Instructions are sent out from the nucleus into the cell. These instructions are made from **mRNA**. The 'm' stands for 'messenger' because it carries a message.

> **KEY POINT**
> **Differences between DNA and mRNA:**
> - RNA has a four-letter code but T is replaced by U.
> - DNA has two strands but mRNA has only one.
> - The mRNA is a copy of a single gene.

When an architect designs a building he does not give his original plans to the builder. He gives the builder a copy and keeps his original plans safe and out of harm's way.

Remember: one gene = one instruction = makes one protein.

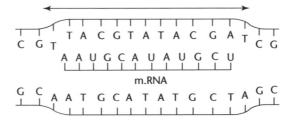

GENE. (Usually contains several hundred bases)

Fig. 11.2

Remember: proteins are made from up to twenty, different amino acids. This means that there must be at least twenty different words.

Once the mRNA has left the nucleus, it lines itself up on a structure called a **ribosome**.

The ribosome is like a 'tape head' in a video player. It reads the message.

The message consists of words, all of which have three 'letters' or 'bases'. Each word codes for a different amino acid. Unlike the English language, there are no spaces between the words.

The protein is put together by another kind of RNA called **tRNA**. The 't' stands for 'transfer', because it transfers amino acids to the ribosome. There is a different kind of tRNA for each different amino acid. The tRNA attaches to its amino acid and when the correct three letter sequence reaches the ribosome, the tRNA connects the amino acids to make the protein.

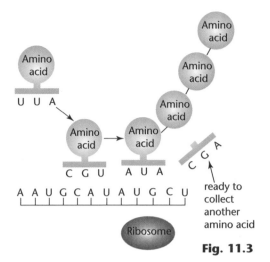

Fig. 11.3

PROGRESS CHECK

1. State which base pairs with C and which base pairs with T.
2. State how many bases code for one amino acid.
3. DNA does not leave the nucleus. State how it gets the instructions on how to make proteins, out into the cell.
4. State which type of RNA carries the amino acid to the ribosome.
5. If a section of DNA has the sequence CGA, state the sequence on the mRNA and the sequence on the tRNA.
6. State two differences between DNA and RNA.
7. State the two functions of DNA.
8. DNA has two strands. Explain why the 'message' is carried by only one of the strands.

1. G and A;   2. Three;   3. Makes a mRNA copy of itself;   4. tRNA;   5. GCU and CGA;   6. RNA has U instead of T and it is single, not double stranded;   7. Store information/ instructions and to make copies of itself;   8. Only the 'sense' strand carries the 'message'. The other strand contains the complementary bases and has a different sequence.

Mutations

Mutations are changes that happen to the DNA. There are various types of mutation, but the one thing they have in common is that they disrupt the message that the DNA is carrying.

Type of mutation	Explanation	Effect
Deletion	A base or sequence of bases is removed	Because each 'word' has three letters, the bases for the gene will be out of sequence. ... ATC GTA CCG ATA TAC ... ... ATC GTAC CGA TAT AC ...
Insertion	A base is added.	The bases will be out of sequence. ... ATC GTA CCG ATA TAC ... ... ATG CGT ACC GAT ATA C ...
Inversion	A sequence of bases is turned round.	Only part of the gene is disrupted. ... ATC GTA CCG ATA TAC ... ... ATC GTA GCC ATA TAC ...
Translocation	Part of the gene is removed and added elsewhere.	This alters the sequence for most of the gene. ... ATC GTA CCG ATA TAC ... ... ATC CCG ATA TAC GTA ...

> **KEY POINT**
> Mutations are caused by chemicals in the environment or radiation. A change to a single base in the human haemoglobin gene can cause sickle cell anaemia.

Gene detection

AQA
Edexcel A   Edexcel B
OCR A ᴬ   OCR A ᴮ
NICCEA
WJEC

Using *in vitro* fertilisation or IVF (this means the sperm and ova are fertilised in a test-tube) it has now becoming possible to ensure that a baby is born without a particular genetic disease. This is possible because modern genetics can detect a faulty gene from a single cell.

Making 'test-tube babies' involves taking eggs from mum and sperm from dad and fertilising together in a test tube. The embryo is then implanted in the mother's womb and grows into a baby.

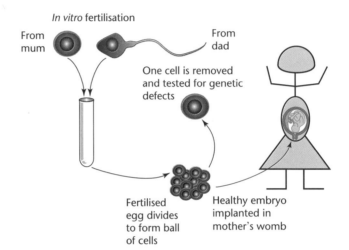

Fig. 11.4

Finding a faulty gene with a gene probe

To find a faulty gene requires a large amount of DNA to be examined. In 1980 it cost £60 to examine each pair of bases. The cost is now less than one pence.

One method for finding a faulty gene is to use a **gene probe**. A gene probe consists of a complementary strand of DNA to which a marker is attached. The patient's DNA is separated into a single strand and the probe then attaches itself to the faulty gene. The presence of the marker tells the scientist that the gene is faulty.

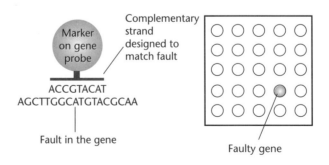

Fig. 11.5

> **PROGRESS CHECK**
>
> 1. Name two different kinds of gene mutation.
> 2. Explain why an insertion or deletion, disrupts the whole gene.
> 3. When testing for a faulty gene, state why the patient's DNA must be separated into a single strand.
>
> 1. Any two from: insertion, deletion, inversion or translocation;   2. All of the bases are moved along thus disrupting each triplet of three bases;   3. A single strand is required so that the complementary strand on the gene probe can match it.

Genetic engineering

AQA
Edexcel A   Edexcel B
OCR A^A   OCR A^B
NICCEA
WJEC

> **KEY POINT**
> Genetic engineering involves taking DNA from one organism and inserting it into the chromosomes of another organism.

The new organism then carries out the instructions on the new DNA. It works because all living organisms use the same code of four bases A, T, C and G in their DNA.

In other words *they all talk the same language.*

> This means that a gene for making vitamin A in a carrot, will still make vitamin A if transferred to rice.

Many people in Asia eat mainly rice. Rice does not normally contain vitamin A. Lack of vitamin A causes damage to their eyes. Scientists have used genetic engineering to make a variety of rice that now makes vitamin A.

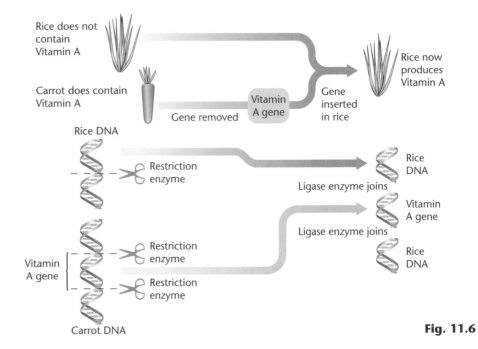

Fig. 11.6

The gene can be removed from the DNA by using **restriction enzymes** that cut the gene out of the DNA. The same enzymes can then be used to cut open the DNA of the new host organism. The new gene is inserted and another enzyme called a **ligase** is used to join the DNA back together.

> **KEY POINT**
> **Genetically modified foods** or **GM foods**, are foods that have been genetically engineered by having a gene from another organism inserted. In other words, their DNA has been modified.

The rice, that can now produce vitamin A and save the sight of millions of people, is an example of a genetically modified food.

Another example of GM food is 'GM soya bean'. The soya bean has been modified to be resistant to a special type of weed-killer. This enables farmers to spray the soya bean with weed-killer to kill weeds without harming the soya bean. It also means that agro-chemical companies make large profits as the farmers not only buy the seed from them, but the weed-killer as well.

> This makes it easier for the farmer but may encourage a greater use of weed-killer.

The ethics of genetic engineering

> Some people are in favour of genetic engineering. They think the benefits will be enormous and worth the risks involved.
>
> Some people are against genetic engineering. They think the risks are too great and that we should not 'play around with nature'.

Benefits:

- cures for diseases, such as cystic fibrosis and cancer

- food which is healthier, stays fresh longer and tastes better.

Risks:

- unknown effects of moving genes from one organism to the other

- new dangerous diseases being created

- against God and nature.

> This is part of the syllabus where you could be asked 'Ideas and Evidence' type questions, i.e. *What do you think and why?*

Genetic fingerprints

AQA
Edexcel A   Edexcel B
OCR A ᴬ   OCR A ᴮ
NICCEA
WJEC

90% of all our DNA is not required to build and maintain our bodies. This surplus DNA is sometimes called 'junk DNA'. However, it can be very useful for making 'genetic fingerprints'.

> The police use 'genetic fingerprints' to identify and convict suspects guilty of a crime.

The junk DNA consists of short sequences that are repeated many times. The pattern of these sequences are unique to a single individual. It is this pattern that forms the 'genetic fingerprint'. It can be used to identify a single individual. Identical twins are the only individuals who have the same genetic fingerprint.

Step 1 – Restriction enzymes cut the DNA into small fragments.

Step 2 – The DNA fragments are placed on a gel. An electric field is applied.

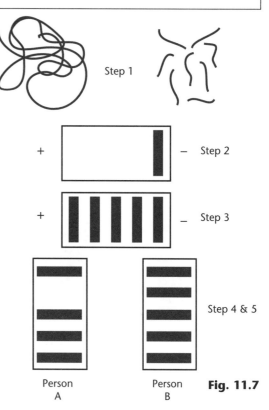

Person A   Person B   **Fig. 11.7**

Step 3 – The fragments separate because the smaller the fragment, the quicker it moves towards the positive electrode.

Step 4 – The fragments are then marked with a gene probe so that they become visible.

Step 5 – The pattern of fragments between two or more individuals can then be compared.

Cloning

AQA
Edexcel A   Edexcel B
OCR A ^A   OCR A ^B
NICCEA
WJEC

A clone is an organism that is genetically identical to another organism. That means that identical twins are clones.

Clones occur in nature all of the time. When you buy a kilogram of apples, they are all genetically identical and are clones. The same applies to any fruit that has been harvested from the same variety of plant, that has been produced by asexual means.

The first mammal to be cloned was 'Dolly' the sheep.

- An ovum is selected from a female sheep and the nucleus is removed from the ovum.

- The DNA is then taken from any cell of another sheep.

- This DNA is then injected into the empty ovum.

- The ovum is returned to the female sheep where it grows and develops in the womb.

Cloning is a technique that could be used for saving species, such as the giant panda and Indian tiger, that are close to extinction.

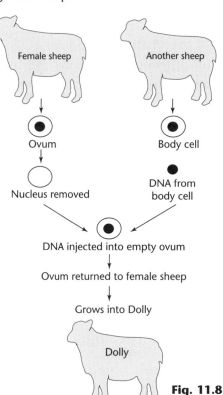

Fig. 11.8

PROGRESS CHECK

1.   State one example of useful genetic engineering.
2.   Name the enzymes that cut and join DNA together.
3.   State one example of genetic engineering that might not be in the consumer's interest.
4.   State which type of individuals will have the same genetic fingerprint.
5.   State one advantage of cloning.
6.   Explain how the process of cloning is carried out.

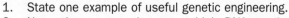

1. Insertion of the vitamin A making gene into rice to prevent eye diseases in people living in Asia;   2. Restriction enzymes cut and ligase enzymes join;   3. Making soya beans resistant to weed killer;   4. Identical twins;   5. Saving endangered species;   6. An ovum is selected from a female sheep and the nucleus is removed from the ovum. The DNA is then taken from any cell of another sheep. This DNA is then injected into the empty ovum. The ovum is returned to the female sheep where it grows and develops in the womb.

Genetics and genetic engineering

Sample GCSE question

1. Richard's mother tells him it is time to have his hair cut. Richard knows that his hair has grown because the cells that make the hair are continuously dividing and copying the DNA in the nucleus.

(a) State one other job carried out by the DNA. [1]

Coding information ✓.

(b) Explain how the DNA in Richard's hair cells makes the protein keratin from which his hair is made. [5]

The two strands of DNA separate ✓ at the keratin gene ✓.
A single strand mRNA copy of the gene is made ✓.
The mRNA leaves the nucleus ✓ and lines up on a ribosome ✓.
Three bases code for one amino acid ✓ and match with a single tRNA ✓. tRNA carry amino acids to the ribosome ✓.
The amino acids link up to form protein ✓.

> This is an example of where there are five marks for the question, but there are more than five opportunities to gain the marks.

> This question is an example of 'continuous prose' where you are expected to write several logical sentences.

(c) Richard is an identical twin. State what this tells you about the DNA of Richard and his twin. [1]

They have identical DNA ✓.

(d) Richard knows that identical twins are also clones.

Dolly the sheep was the first cloned mammal to be made in the laboratory. Explain why Dolly is referred to as a clone. [2]

Dolly has identical DNA ✓ to the original sheep from which the DNA was taken ✓.

(e) Explain the steps that would be taken by the scientists, to clone another sheep from Dolly. [4]

> This is a tricky question as it asks you not to describe how Dolly was made, but to use that information to explain how to make another clone of Dolly.

DNA removed from the ova of another sheep ✓. DNA taken from any cell from Dolly ✓.
Dolly's DNA is then inserted into the other sheep's ovum ✓.
The ovum is then replaced in the womb of the sheep to grow ✓.

(f) Explain with reasons whether you think cloning animals is a good idea, or whether the process should be banned. [3]

> This is an example of an Ideas and Evidence question.

For:
Rare and endangered species could be saved from extinction, or any other good reason ✓.

Against:
There may be unknown effects in the offspring. This could have serious consequences in humans ✓.
The decision with a reason ✓.

> The mark for making your decision will only be awarded if you give a reason for your decision. There is no right or wrong decision for this question.

Exam practice questions

1. The following strand of DNA is part of a gene.

-A-C-C-G-T-A-C-T-G-G-A-C-A-

(a) Write down the sequence of bases that will make up the complementary strand of mRNA. **[3]**

(b) The mRNA leaves the nucleus and lines itself up on the ribosome.

State the base sequence of the first two tRNA molecules that each code for one amino acid. **[3]**

(c) A mutation occurs to the original DNA strand. The first base A, is removed and replaced by a G.

(i) State two possible causes of this mutation. **[2]**

(ii) State the name of this type of mutation. **[1]**

(iii) Explain what effect this mutation would have on the expression of the gene. **[2]**

The following table lists which three bases on the mRNA molecule, codes for which amino acid.

mRNA	Amino acid
GUG	Valine
UGU	Cysteine
GGG	Glycine
CGC	Arginine

(d) State which amino acid is coded, by the last three bases, ACA, in the DNA strand. **[1]**

2. Mary and David have a child with cystic fibrosis. They know that this condition is inherited. They decide to have another child, but do not want it to suffer from cystic fibrosis. After talking to the doctor they decide to have *in vitro* fertilisation.

(a) State what *in vitro* means. **[1]**

(b) The doctor tells them that after the sperm has fertilised the egg and the egg has started to divide, he will remove one cell for genetic testing.

State what effect this removal will have on the rest of the cells. **[1]**

(c) The cell is tested to see if it contains the gene for cystic fibrosis.

State what the doctor will be able to tell the parents, if the gene is present in the cell. **[1]**

(d) When testing for the gene the doctor separates the DNA in the cell into a single strand.

Explain why the doctor does this. **[2]**

(e) He mixes the single strand with complementary DNA for the cystic fibrosis gene.

Explain what will happen to this complementary DNA. **[2]**

(f) The complementary DNA strand has a marker attached to it. When he examines the results of the test he cannot see any markers. State what this tells you about Mary and David's embryo. **[1]**

Further physiology

The following topics are covered in this section:

- ● **Food and feeding**   ● **Excretion**
- ● **Coordination and movement**

LEARNING SUMMARY

After studying this section you should be able to:

- ● *identify the components of a balanced diet and explain their roles*
- ● *explain how certain diseases are caused by the lack of a balanced diet*
- ● *realise that dietary requirements vary in different people*
- ● *understand how the teeth and digestive systems of animals are adapted to different diets*
- ● *describe the functions of kidney tubules and how kidney failure is treated*
- ● *describe how the skeleton of mammals allows movement*
- ● *explain how the body adapts to the needs of exercise*
- ● *locate the coordination of different parts of the body to different areas of the brain.*

KEY POINT

Physiology is the study of the living processes in organisms and how they respond to changing conditions. This chapter concentrates on animal physiology and builds on some of the topics that were covered in Chapter 2 Humans as organisms.

Food and feeding

AQA
OCR A ^A^
NICCEA

A balanced diet

The body needs the correct combination of seven different types of food substance in the diet in order to remain healthy. These substances must be combined in the correct amounts.

Remember: a balanced diet is not 'enough' of each type of substance but the correct amount.

KEY POINT

The intake of the correct amounts of these food substances is called a balanced diet.

Different food substances are needed in the body for different functions and are obtained from different foods. A mixture of different foods is therefore needed in a balanced diet. The functions of many of these food substances have been discovered by looking at what happens if the body is lacking the substance.

KEY POINT

A disease caused by the lack of a food substance is called a deficiency disease.

Food substance	Use in the body	Deficiency disease	Good food source
protein	growth and repair of cells, making enzymes	kwashiorkor: swollen abdomen, loss of hair	meat, fish, soya
digestible carbohydrate	supply of energy (glycogen is a store of energy, sugars are a ready supply)	lack of carbohydrate is often linked to lack of enough food of any type, i.e. starvation.	rice, potatoes, cereals
fats	store of energy	lack of certain fatty acids may cause various diseases	dairy produce, meat, fish, nuts
vitamins A C D	to make light sensitive chemical in rods to make connective tissue to absorb enough calcium from the intestine	poor night vision scurvy: poor healing of wounds and bleeding gums rickets: weak bones	cod liver oil, butter citrus fruit, fresh vegetables dairy produce (can be made by sunlight on the skin)
minerals iron calcium	to produce haemoglobin for red blood cells strengthening bones and teeth	anaemia: lack of red blood cells rickets and poorly developed teeth	liver, egg yolk dairy produce, bread
fibre	to allow the correct rate of peristalsis	constipation, appendicitis and bowel cancer	wholemeal bread, fruit

In western society it is often more common to find diseases that are caused by too much of a substance than to find deficiency diseases. Examples are:

Too much fat (particularly animal fat) can cause blockages of the coronary arteries of the heart leading to **coronary heart disease.**	Too much sugar in the diet may lead to acid production by bacteria in the mouth leading to **tooth decay.**	Too much salt in the diet may lead to increased blood volume and so increased **blood pressure.**

Dietary requirements

The amounts of the seven different food substances that are needed to make a balanced diet depend on a number of factors. These include:

- *age* – babies have large requirements for most substances compared to their size because they are growing. This growth slows but then adolescents will need large amounts of proteins and other nutrients as they undergo a growth spurt

Girls often undergo an adolescent growth spurt before boys.

- *sex* – women need more iron to replace blood lost in menstruation, there are various differences in requirements due to hormonal differences

- *level of physical activity* – people with high levels of physical activity will need large amounts of energy rich foods in their diet

Breast feeding will also increase demands.

- *pregnancy* – pregnant women need extra nutrients to supply the baby with its requirements for growth.

Different types of feeding in mammals

Mammals have teeth which are often used to bite food and chew it into small pieces. Teeth have different shapes and this makes them suited to a particular function. The selection of teeth that an animal possesses depends on the type of food that it eats.

> **KEY POINT** Humans have four types of teeth. This allows them to eat a range of plant and animal material.

Remember: breaking food down into smaller pieces gives enzymes a larger surface area to attack.

Premolars may have one or two roots, molars have two or three.

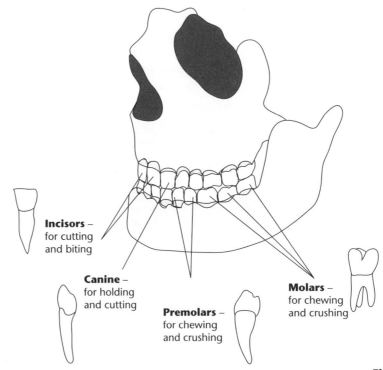

Incisors – for cutting and biting

Canine – for holding and cutting

Premolars – for chewing and crushing

Molars – for chewing and crushing

Fig. 12.1

Other animals may have different combinations of these types of teeth and this enables them to deal with different diets. There are also differences in their skulls and jaws.

Meat eaters, such as the dog, are called carnivores and plant eaters are herbivores.

> **KEY POINT** Dogs have teeth and jaws that are adapted for eating meat. Other animals, such as sheep, are adapted for eating plant material.

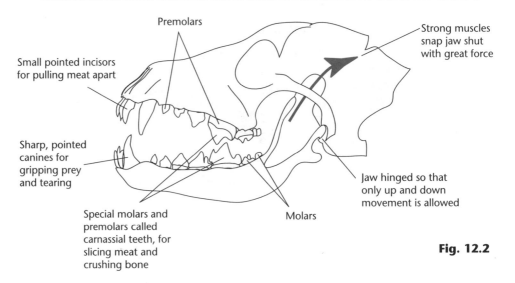

Premolars

Small pointed incisors for pulling meat apart

Sharp, pointed canines for gripping prey and tearing

Special molars and premolars called carnassial teeth, for slicing meat and crushing bone

Molars

Strong muscles snap jaw shut with great force

Jaw hinged so that only up and down movement is allowed

Fig. 12.2

As well as differences in their teeth, herbivores and carnivores have differences in their digestive systems. The main problem for herbivores is cellulose:

Mammals cannot make an enzyme to digest cellulose.

KEY POINT Herbivores have to keep cellulose digesting bacteria in their gut to break down the cellulose into sugars.

Plant material is made of plant cells, all of which are surrounded by a cell wall made of cellulose!

Different herbivores keep these bacteria in different parts of their digestive system. In sheep and cows they are in a special sac called the rumen. This is between the oesophagus and the true stomach. In rabbits, the bacteria are in the appendix, which is between the small and large intestine:

Animals like cows and sheep are called ruminants because of their rumen.

This is an example of mutualism because the bacteria get a supply of food and the herbivore gets its cellulose digested.

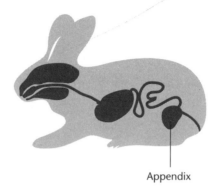

Appendix

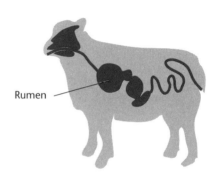

Rumen

As the appendix is after the site of absorption (small intestine), rabbits need to eat their faeces so that the sugars can be absorbed!

To help the bacteria digest the cellulose, the sheep brings up partly digested food to give it some extra chewing in the mouth.

Fig. 12.3

Carnivores do not need cellulose digesting bacteria. Their digestive system is often shorter than herbivores because meat is easier to digest than the much tougher plant material.

PROGRESS CHECK

1. Why are people with anaemia often tired and short of breath?
2. Why do people with kwashiorkor often lose their hair?
3. Why do women need more calcium when they are pregnant?
4. Sheep do not have canine teeth. Why is this?
5. Why does rechewing food help the microbes in a sheep's rumen to digest the cellulose faster?

1. Anaemia is a lack of red blood cells so sufferers cannot transport enough oxygen to the tissues in order to carry out enough respiration; 2. Hair is made largely of protein and kwashiorkor sufferers lack protein; 3. Pregnant women need to supply the fetus with enough calcium for its growing bones and teeth; 4. Canines are for tearing meat and sheep are herbivores; 5. Chewing provides the food with a larger surface area for the enzymes to act on.

Excretion

AQA

OCR A ᴬ

NICCEA

WJEC

Kidney tubules

Chapter 2 explains how the kidney works by filtering the blood under pressure and then reabsorbing the useful substances back into the blood. This all occurs in long tubules called nephrons. The diagram on page 38 shows the first part of a nephron but a full nephron is shown here:

> If the blood pressure drops too much then the filtration may stop.

> Selective reabsorption of glucose and amino acids is by active transport.

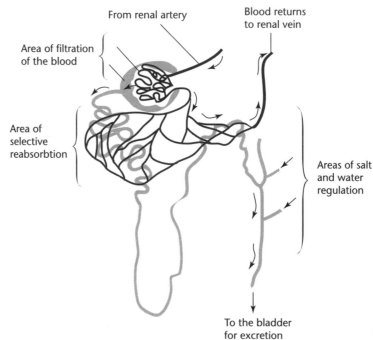

From renal artery

Blood returns to renal vein

Area of filtration of the blood

Area of selective reabsorbtion

Areas of salt and water regulation

To the bladder for excretion

Fig. 12.4

> **KEY POINT** The nephron has three main areas, a filter unit, an area of selective reabsorption and an area for salt and water regulation.

How the nephron changes the composition of the fluid is shown in the table:

> Exam questions often ask for an explanation of the zero figures in this table.

substance	Percentage of substance in		
	blood plasma	filtered fluid	urine
water	90	99	97
proteins	9	0	0
glucose	0.1	0.1	0
urea	0.03	0.03	2.0

- Protein is absent from the urine because the molecules are too large to pass through the filter unit and so stay in the blood.
- Glucose is absent from the urine because it is selectively reabsorbed from the tubule back into the blood.

Treating kidney disease

People may suffer from kidney failure for a number of reasons. A person can survive if half of their nephrons are still working but if the situation worsens then there are two options:

> A person can therefore survive with one working kidney.

Fig. 12.5

Kidney failure

Kidney dialysis   Kidney transplant

Kidney dialysis

> **KEY POINT** **Kidney dialysis involves linking the person up to a dialysis machine. This takes over the role of the kidneys and removes waste substances from the blood.**

> The anticoagulant stops the blood clotting while it is in the machine.

> The bubble trap stops bubbles of air entering the body in the blood.

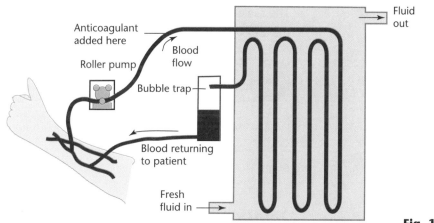

Fig. 12.6

The blood is removed from a vein in the arm. It then passes through a long coiled tube made of partially permeable cellophane. The fluid surrounding the tube contains water, salts, glucose and amino acids but no waste materials, such as urea. These waste materials therefore diffuse out of the blood into the fluid.

Transplants

> Over 2000 kidney transplants are carried out in the UK every year but donors are in short supply.

As a person can survive with one kidney, it is possible for a person to donate one kidney to be transplanted into another person.

> **KEY POINT** **The main problem with transplants is to prevent the person's immune system rejecting the transplanted kidney.**

This is avoided by taking certain precautions:

- making sure that the donor has a similar 'tissue type' to the patient
- treating the patient with drugs or radiation to make their immune system less effective.

Coordination and movement

Muscles, joints and movement

The human skeleton is contained inside the body and carries out a number of important functions:

- support for the body

- protection for vital organs, such as the brain

- movement.

The skeleton is made up largely of bone and it is the structure of bone that makes these functions possible.

Look at the effect of lack of vitamin D in Chapter 12, page 141.

> **KEY POINT**
> Bone is hardened by calcium phosphate but contains living cells and protein, which stop it from being brittle.

Where two bones meet they form a joint. Some of these joints are fused, such as the bones in the skull, but others are moveable joints. These joints act as pivots across which muscles are attached. It is this that allows movement.

Joints

Arthritis is a common disease of joints, which is sometimes caused by the cartilage breaking down.

> **KEY POINT**
> Moveable or synovial joints are specially adapted to allow smooth movement and to absorb shock.

Synovial fluid is a slimy liquid that helps to reduce friction.

Bone

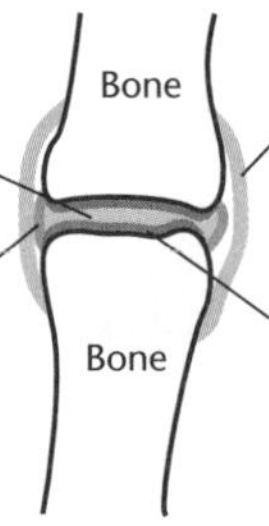

Ligaments connect bone to bone. They are strong but elastic, absorbing movement but preventing dislocation.

Synovial membrane produces synovial fluid.

Bone

Cartilage coats the ends of the bones and is smooth and rubbery. This reduces friction and helps to absorb shock.

Fig. 12.7

The use of pain killing sprays may allow more damage to occur because they allow further exercise to take place after injury.

Although the tissues in the joint are well adapted for their functions, care must be taken when exercising to avoid:

Dislocations

This occurs when a bone is forced out of position in a joint.

Sprains

This occurs when the ligaments and other tissues are torn by a sudden wrench.

Fig. 12.8

Muscles

Muscles are the main effector organs in the body. They contain muscle fibres that can shorten and so make the muscle contract. In order to occur, this needs energy from respiration. Muscles can only contract and pull on bones but cannot actively expand or push.

> **KEY POINT** Muscles therefore have to be arranged in pairs called antagonistic pairs. When one contracts the other relaxes and vice versa.

The biceps and the triceps are the antagonistic muscles in the human arm:

Make sure that you can predict the effect of muscle contraction if you were given a diagram of a different joint.

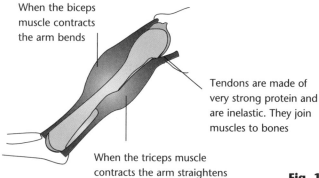

When the biceps muscle contracts the arm bends

Tendons are made of very strong protein and are inelastic. They join muscles to bones

When the triceps muscle contracts the arm straightens

Fig. 12.9

Regular exercise will help muscles to work efficiently in a number of ways by:

- keeping muscles in a slightly contracted state (tone) so that they are ready to contract
- increasing the strength of contraction and preventing aches after exercise
- keeping joints working smoothly
- improving the blood supply to the muscles and the efficiency of the heart and lungs.

The circulation and exercise

The circulation needs to adapt and change depending on the level of activity of the body. During exercise the muscles must be supplied with more oxygen and waste products, such as carbon dioxide, must be removed.

> **KEY POINT** Adjustments in the activity of the heart and the breathing system are brought about by nerves and hormones.

Blood pressure and heart rate

It is vital to keep blood pressure constant for two main reasons:

 An increase in blood pressure may lead to strokes or heart disease.

 A low blood pressure may lead to kidney failure.

Fig. 12.10

This type of control is called negative feedback.

Pressure receptors in the walls of certain arteries detect the pressure of the blood. If the pressure rises then nerve impulses are sent to the brain. The brain will then send impulses to the heart to slow down the heart rate and so reduce the blood pressure.

Preparing for exercise

Even before a person starts any exercise, changes already occur in the body. These are mainly due to the hormone **adrenaline**.

 KEY POINT The hormone adrenaline brings about a number of changes in the body that prepare the person for increased activity.

Adrenaline is sometimes called the 'flight or fight' hormone.

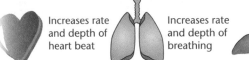

Increases rate and depth of heart beat

Increases rate and depth of breathing

Converts glycogen to glucose in the liver

Increases blood supply to the muscles and decreases supply to the skin and gut

Fig. 12.11

Changes during exercise

As the person starts to exercise, the muscles start to contract more and produce more **carbon dioxide** and **lactic acid**. They also use up more **oxygen**. The lack of carbon dioxide is detected in the brain and more adrenaline is made. Nerve impulses are also sent to the lungs and heart and have the same effect as adrenaline.

The graphs show some of the changes that typically take place:

The increase in heart rate will also increase the blood pressure but the high levels of carbon dioxide take priority and so the heart rate is not reduced.

Once the exercise stops, carbon dioxide levels can return to normal and the lactic acid can be broken down. The brain now returns the heart rate and blood pressure to normal.

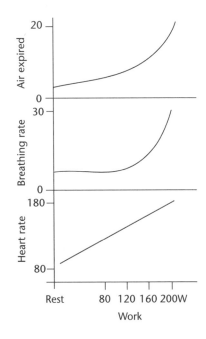

Fig. 12.12

Coordination and the brain

The control of movement and the coordination of the changes that occur during exercise all occur in the brain.

The pathway for the withdrawal reflex is shown on page 34.

Some simple reflexes, such as the withdrawal reflex, pass via the spinal cord and do not involve the brain. Other reflexes, such as blinking, involve the brain but do not involve conscious thought.

KEY POINT Different areas of the brain are responsible for different functions.

- **Medulla** – this is the part of the brain that joins with the spinal cord. Blood pressure, and heart and breathing rate are controlled here.

The function of many of these parts has been discovered by studying the effects of injury.

- **Cerebellum** – this is an outgrowth from the back of the brain that controls balance and coordinates movement.

- **Cerebrum** – the outer layer is called the cerebral cortex. This area is responsible for voluntary actions.

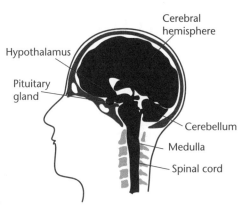

Fig. 12.13

Voluntary actions

> **KEY POINT** A voluntary action involves conscious thought.

The cerebral cortex is divided up into different areas.

The sensory cortex receives information from the sense organs. Other areas in the front of the cortex analyse the information and then decide on a response, using intelligence and memory areas. The motor area then initiates a response. The nerve impulses are then sent down the spinal cord to the effectors, via motor neurones.

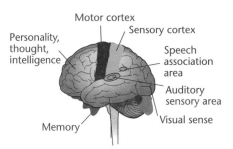

Fig. 12.14

Conditioned reflexes

Many of our responses, such as walking, are thought to involve this type of reflex.

> **KEY POINT** A conditioned reflex is an automatic response to a stimulus that has been learnt.

A Russian scientist called **Ivan Pavlov** carried out some experiments on dogs. He rang a bell immediately prior to feeding the dogs. He found that after a while, the dogs would produce saliva when the bell rang, even without food being provided. He called this type of 'learnt' response, that does not involve conscious thought, a **conditioned reflex**.

PROGRESS CHECK

1. Why are some moveable joints called synovial joints?
2. What is the difference between ligaments and tendons?
3. Name one pair of antagonistic muscles.
4. What makes an athlete's heart rate increase before he starts running a race?
5. What areas of the body have the largest input to the sensory cortex?

1. They contain a fluid for lubrication called synovial fluid; 2. Ligaments are elastic and join bone to bone at joints, tendons are inelastic and join muscle to bone; 3. Biceps and triceps; 4. Adrenaline production; 5. Areas with many sensory receptors such as the fingers, lips and tongue.

Sample GCSE question

1. The amount of each food substance needed by the body varies.

   The following table shows the recommended daily intake of four substances for different aged females.

age	protein (g)	calcium (mg)	iron (mg)	vitamin C (mg)	
3 months	12.5	525	1.7	25	
6 years	19.7	450	6.1	30	
14 years	41.2	800	14.8	35	
20 years	45.0	700	14.8	40	

   Use this table and your biological knowledge to answer the questions that follow.

(a) (i) What may happen to a 20 year old who takes in less than 40 mg of vitamin C per day over a long period of time? **[2]**

> There are 2 marks here so the examiner does not just want the name of the disease.

> *The person may suffer from scurvy ✓ with bleeding gums ✓.*

   (ii) Why is more calcium needed each day at 3 months than at six years? **[2]**

> *At 3 months the baby is growing faster, ✓ calcium is needed for the growth of bones ✓ and teeth ✓.*

   (iii) A 14 year old boy is recommended to take in 11.3 mg of iron a day. Suggest why this is different to the 14 year old girl's requirements. **[2]**

> *The girl loses blood monthly during menstruation ✓.*
> *More iron is needed to make haemoglobin to replace the loss ✓.*

> Menstruation is the medical name for periods.

(b) The following is the contents from a packet of biscuits.

Contents per 100 g	
protein	9.0 g
carbohydrate	68.7 g
fat	14.0 g

   (i) A 20 year-old woman eats 200 g of the biscuits. What percentage of her daily protein requirement has she eaten? **[3]**

> *200 g supply 9.0 × 2 = 18 g ✓*
> *percentage supplied · 18/45 × 100 ✓ = 40% ✓*

> Remember that food packets usually give the contents in 100 g of the food.

   (ii) The biscuits are advertised as being low in salt. Why is this good for the body? **[2]**

> *High salt levels may cause an increase in blood pressure ✓. High blood pressure increases the risk of strokes or heart disease ✓.*

Exam practice questions

1. The diagram shows a section through the hip joint.

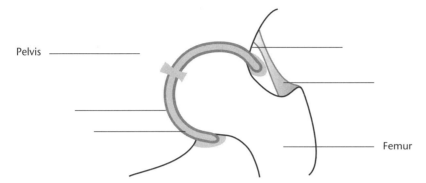

Pelvis

Femur

(a) Finish the diagram by adding the four missing labels. **[4]**

(b) Write down the name of each of these four skeletal tissues by the side of the correct description. **[4]**

An elastic tissue that joins bone to bone _____

A hard but not brittle tissue that contains living cells _____

A rubbery tissue that contains living cells _____

A strong inelastic tissue that joins muscle to bone _____

(c) Explain why muscles are always found in pairs in the body. **[3]**

2. The diagram shows the outline of a section through the human brain.

(a) Write the letters A,B,C on the diagram to locate the following: **[3]**

(i) the area responsible for controlling balance (mark A)

(ii) the area that controls blood pressure (mark B)

(iii) the area that initiates a voluntary action (mark C)

(b) Explain how the body keeps blood pressure within narrow limits **[4]**

(c) During exercise, the body allows the blood pressure to rise.

Explain why this is necessary. **[3]**

Food production

The following topics are covered in this section:

- World food shortage
- Plant disease and its control
- Agriculture
- Hormones and food production

LEARNING SUMMARY

After studying this section you should be able to:

- understand biotechnology's potential to solve the world food shortage problem
- recall how pH affects soils
- appreciate how hydroponics and greenhouses can increase food production
- understand the effects of using fertilisers
- appreciate the importance of weed control
- state an example of and understand plant disease control.

KEY POINT

Many parts of the world are suffering with a severe shortage of food. This unit considers this problem and examines various ways in which this shortage of food can be overcome.
Science is rapidly developing new methods of food production that enable a larger amount of food to be produced using less space and less time with less effort.

World food shortage

AQA
Edexcel A   Edexcel B
OCR A ᴬ   OCR A ᴮ
NICCEA
WJEC

The problem with food production is that large scale efficient farming is already happening in the developed countries where there is plenty of food, but it is not happening in the underdeveloped countries where there is a food shortage. Food production in underdeveloped countries can also be low due to the poor climate, i.e. some countries are too hot or cold, or have too little water.

Biotechnology can help

Fermentation
Microorganisms, such as fungi, can be grown in fermenters which use a small amount of space, to quickly produce high protein foods like mycoprotein.

Increased food production

Gene technology
Genetic modification of plants to produce pest resistant varieties with a longer shelf life and better nutritional qualities. They can also be made to withstand a greater environmental range.

Manipulating reproduction
Using tissue culture (micropropagation) to produce virus free, high quality plants.

Fig. 13.1

Agriculture

AQA
Edexcel A Edexcel B
OCR A ᴬ OCR A ᴮ
NICCEA
WJEC

Soil and pH

Different kinds of soils have a different pH. The pH of a soil is important because different types of crops require different kinds of pH in the soil.

If the soil is too acid, the farmer can add lime to neutralise it.

If the soil is too alkaline, the farmer can add peat to neutralise it.

> The farmer needs to understand his soil type so that he knows which crops will grow well.

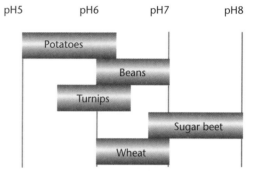

Fig. 13.2

> **KEY POINT** Most crops grow best in a soil that is slightly alkaline.

Hydroponics and greenhouses

Increased food production can be achieved by controlling the environment.

> Farmers need to calculate whether the expense of building the greenhouse will be repaid by extra crop production.

Greenhouses allow the farmer to control the growing environment by controlling temperature, water and nutrients. They also prevent many pests from getting to the crop.

Greenhouses grow more food per unit area.

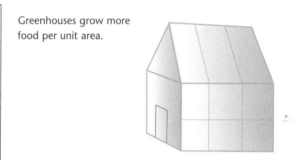

Fig. 13.3

Hydroponics can be used in the greenhouse to give the farmer even greater control over the water and nutrients that the plants receive.

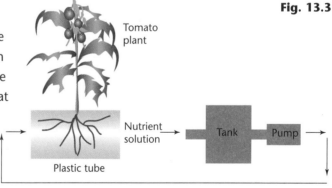

Fig. 13.4

> **KEY POINT** Hydroponics means growing plants without soil.

In the 'nutrient film' technique the plants are grown in plastic tubes through which a nutrient solution passes. The solution is pumped round the tubes from a large storage tank.

Use of fertilisers

Modern intensive farming methods only work because of the use of fertilisers.

Growing crops in the same field for year after year will inevitably remove nutrients from the soil. These have to be replaced by using fertilisers. The use of fertilisers, particularly nitrates, leads to much greater crop production.

> **KEY POINT**
> **Most fertilisers contain nitrates, phosphates and potassium. They are known as NPK fertilisers.**

However their use does have its disadvantages:

- **soil structure** – organic fertilisers, such as manure, adds humus to the soil. This gives the soil a good crumb structure. Inorganic fertilisers do not add humus and crumb structure deteriorates.

- **pollution of rivers** – the mineral salts are often washed out of the soil and into rivers. This can lead to **eutrophication**.

Eutrophication causes the death of fish due to oxygen shortage. This is how it works:

1. Fertilisers get leached or washed into the river.

2. Fertilisers cause a rapid growth of algae in the river.

3. Only algae near the surface get light and survive. The rest of the algae are starved of light and die.

4. Bacteria grow on the dead algae and cause them to rot.

5. This leads to a rapid growth in the number of bacteria.

6. The bacteria need oxygen to grow and use it all up.

7. Fish do not have enough oxygen and die.

> Candidates usually do not do well on questions about eutrophication. You should learn these seven key points.

> **PROGRESS CHECK**
>
> 1. Describe three ways that biotechnology can improve crop production.
> 2. Give an example of a crop that would grow best on an acid soil and a crop that would grow best on an alkaline soil.
> 3. State two ways that a farmer can increase crop production by controlling the plant's environment.
> 4. State two advantages of using fertilisers.
> 5. State two disadvantages of using fertilisers.
> 6. Explain how using fertilisers can lead to eutrophication.
>
> 1. Using fermentation, gene technology and manipulating reproduction;   2. Acid = potatoes, alkaline = sugar beet;   3. Using greenhouses and hydroponics;   4. Increased crop production and using the same land over and over again;   5. Poor crumb structure and eutrophication;   6. Fertilisers leach or get washed into the river. Fertilisers cause a rapid growth of algae in the river. Only algae near the surface get light and survive. The rest of the algae are starved of light and die. Bacteria grow on the dead algae and cause them to rot. This leads to a rapid growth in the number of bacteria. The bacteria need oxygen to grow and use it all up. Fish do not have enough oxygen and die.

Questions on this section usually involve interpreting evidence from tables of data.

Weed control

KEY POINT A weed is a plant growing in a place where it is not wanted.

- Weeds can be controlled by using mechanical methods, such as rotivators and hoeing.
- Weeds can be controlled by using chemical methods, such as weedkillers.

KEY POINT Mechanical methods involve uprooting the weed and physically destroying it.
Chemical methods involve spraying the weed with a chemical that is absorbed by the weed and kills it.

Modern weedkillers are usually selective. This means that they will kill certain weeds without harming the crops.

Genetic engineering has been used to develop some crop plants that are resistant to weedkillers. This may lead to greater crop yield, but it could also lead to the use of more weedkillers.

Plant disease and its control

AQA
Edexcel A   Edexcel B
OCR A ^A   OCR A ^B
NICCEA
WJEC

Powdery mildew is a fungus that attacks other plants. It is a parasite and lives on cereals such as wheat and barley. Although it normally does not kill the host plant on which it lives, it weakens its and seriously reduces the yield of crop that the farmer can harvest.

This is just an example of a plant disease. It does not matter if your teacher uses a different example.

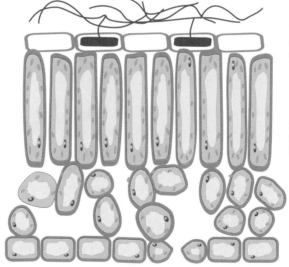

Fungus growing on upper surface of the leaf

Section through leaf

Fig. 13.5

Powdery mildew grows mainly on the upper surface of the leaf, reducing the rate at which the plant can photosynthesise. The fungus produces hyphae that penetrate the upper epidermal cells to absorb food and nutrients, so the plant has less food to store in seeds, which we want to harvest.

Controlling plant diseases

 KEY POINT Plants diseases can be controlled not only by using pesticides and fungicides, but also by using good farming practice.

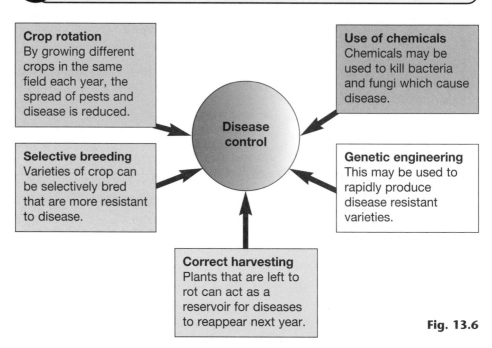

Crop rotation
By growing different crops in the same field each year, the spread of pests and disease is reduced.

Use of chemicals
Chemicals may be used to kill bacteria and fungi which cause disease.

Disease control

Selective breeding
Varieties of crop can be selectively bred that are more resistant to disease.

Genetic engineering
This may be used to rapidly produce disease resistant varieties.

Correct harvesting
Plants that are left to rot can act as a reservoir for diseases to reappear next year.

Fig. 13.6

Hormones and food production

AQA
Edexcel A   Edexcel B
OCR A ᴬ   OCR A ᴮ
NICCEA
WJEC

Hormones are chemical messengers. Some of them control the processes of growth and reproduction. They can be used deliberately to increase meat and milk production in cattle.

In some countries, but not in Britain, **BST** or **bovine somatotrophin** can be injected into cows to increase milk production. The long term effect on humans who drink the milk is still unknown, but it does lead to an increase in infection in the cows.

Fig. 13.7

PROGRESS CHECK

1. State two different methods of weed control.
2. State an example of a plant disease and describe its effect on the plant.
3. Explain how plant diseases may be controlled.
4. State an example of food production being increased by the use of artificial hormones.

1. Chemical and mechanical;   2. Powdery mildew – it weakens the plant and reduces photosynthesis;   3. Crop rotation. Use of chemicals. Selective breeding. Genetic engineering. Correct harvesting;   4. Somatotrophin used to increase milk yield in cows.

Sample GCSE question

1.

(a) The shortage of food is a major problem in some countries.

Questions that start with 'suggest' are ones where you can use your own ideas. The specific answers may not have been taught you at school.

Suggest how each of the following biotechnologies can help to relieve this shortage.

(i) Fermentation [2]

High protein food can be produced quickly ✓ using a small amount of space ✓.

(ii) Gene technology [2]

Foods can be made to be pest resistant ✓ with a longer shelf life ✓.

(iii) Manipulation of reproduction. [2]

Tissue culture (micropropagation) ✓ can be used to quickly produce virus free high quality plants ✓.

(b) Reshma grows crops in her garden. She uses the following table to see which crops grow best at which pH.

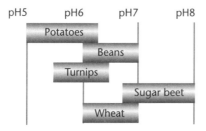

(i) State what is meant by the pH of the soil. [1]

This is a question that requires you to interpret data. It has been made slightly more difficult by asking for pH7.5 which is not shown on the chart.

How acid or alkali the soil is ✓.

(ii) The soil in Reshma's garden is pH7.5. State which crop will grow best. [1]

Sugar beet ✓.

(iii) Reshma wants to grow potatoes. Explain what she must do to the soil so that the potatoes will grow well. [2]

Add peat ✓ to make the soil more acid ✓.

(c) Reshma decides to grow some tomatoes in her greenhouse. Explain why she expects the tomatoes to grow better in the greenhouse, than in the garden. [2]

Controlled environment ✓ with one example, such as temperature ✓. or easier control ✓ of pests ✓.

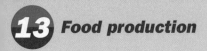

Exam practice questions

1.

(a) Farmer Giles notices that the river near his farm has turned green.

He thinks it may be due to the fertiliser that he has been using on his fields.

    (i) Name the process that may have caused the river to go green. **[1]**

    (ii) Explain how the fertiliser might have got into the river. **[1]**

    (iii) Explain how this process causes large amounts of algae in the river, to die. **[3]**

    (iv) Explain why the death of the algae, cause the fish to die as well. **[2]**

(b) Farmer Giles notices that the fertiliser also makes the weeds in the crop grow.

    (i) Chemicals can be used to control weeds.

       State the name of **one other** method of weed control. **[1]**

    (ii) Farmer Giles decides to use a chemical weed killer.

       Suggest and explain two reasons why the weed-killer kills the weeds but not the crop. **[4]**

(c) The crop gets attacked by Powdery Mildew.

    (i) Explain what this will do to the yield of the crop. **[1]**

    (ii) Explain how each of the following methods may be used to control the disease:

       crop rotation **[1]**

       use of chemicals **[1]**

       selective breeding **[1]**

       genetic engineering. **[1]**

2. Some parts of the world are short of food, causing people to starve.

(a) State two reasons why some parts of the world have plenty of food and others have too little. **[2]**

(b) Explain how each of the following methods can be used to increase food production in parts of the world that have a food shortage.

    (i) Fermentation **[2]**

    (ii) Gene technology **[2]**

    (iii) Manipulating reproduction **[2]**

(c) Increased food production can be achieved by controlling the environment.

Explain how the environment is controlled by each of the following methods.

    (i) Greenhouses **[2]**

    (ii) Hydroponics **[2]**

Further ecology

The following topics are covered in this section:

● *Ecosystems*

After studying this section you should be able to:

LEARNING SUMMARY

● *understand that some ecosystems are natural and some are man-made*
● *use the 'Species Diversity Index'*
● *understand about species conservation*
● *recall and understand how certain pollutants work*
● *understand an 'Environmental Impact Statement'*
● *appreciate how conflicts of interest can arise and be solved.*

> **KEY POINT**
>
> Ecology is the study of the relationship between living things and their environment. This unit looks at how human beings have affected the environment, both by damaging it and also by trying to protect it for future generations.

Ecosystems

AQA
Edexcel A   Edexcel B
OCR A ᴬ   OCR A ᴮ
NICCEA
WJEC

An **ecosystem** is an environment plus the living organisms within that environment.

Fig. 14.1

> **KEY POINT**
>
> Some ecosystems are natural, e.g. deserts, polar regions, oceans, tropical rain forests. Other ecosystems are man-made or artificial, e.g. fields, parks, built up areas.

Some ecosystems that we might think are natural, such as moorland and pine forests, are in fact artificial ecosystems. Pine forests are planted and grown for their wood and moorland is maintained for grouse shooting. A few hundred years ago, over 50% of England was covered by broad-leaf woodland. This figure is now less than 4%. Open hillsides are kept free of trees, by farmers grazing their sheep. If the environment was left alone, after a few years, it would revert back to broad-leaf woodland.

Species diversity index

KEY POINT The variety of different species within an ecosystem is called **diversity. The greater the number of species, the greater the diversity.**

In other words, if one food source disappears, there will be plenty of other food available.

The greater the diversity within an ecosystem, the more stable and secure that ecosystem is.

The food web will be very large so any change to one part of the web will make only a very small change to the whole ecosystem.

KEY POINT The species diversity index, **measures how much diversity there is within an ecosystem.**

Do not worry if your teacher uses a different way of calculating the diversity index.

There are several different ways of calculating this index.

A simple way of calculating a Species Diversity Index is to calculate the **Biotic Index** for a particular type of ecosystem.

The Field Studies Council Biotic Index calculates the amount of pollution in fresh water rivers.

How the Biotic Index works

An exam question could be based on giving you data on a particular river and a copy of the score of each organism. You would then be asked to calculate the Index.

The index works by looking for particular organisms in the river and giving them a score if they are present. The total score is then divided by the number of different species found and this gives a number between 1 and 10. The score of 0 indicates a highly polluted river and a score of 10 indicates a very clean river, such as an unpolluted upland stream.

KEY POINT The greater the diversity, the higher the index, the cleaner the water.

Species conservation – extinction is forever

Some species, such as the giant panda and the Indian tiger, are close to extinction.

Many zoos around the world have started breeding programmes to try to protect endangered species. However, it is only by protecting their natural environment that we can ensure their long term survival. In Great Britain special environments can be protected by declaring them as SSSI or **Sites of Special Scientific Interest**. This protects the sites from further development.

On a small scale, conservation can be carried out by:

● planting trees

● making new ponds and lakes

● putting up bird and bat boxes.

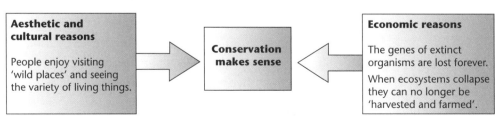

Aesthetic and cultural reasons

People enjoy visiting 'wild places' and seeing the variety of living things.

Conservation makes sense

Economic reasons

The genes of extinct organisms are lost forever.

When ecosystems collapse they can no longer be 'harvested and farmed'.

Fig.14.2

Pollution

Four main causes of pollution are:

- car exhausts
- fossil fuels
- sewage
- fertiliser run-off.

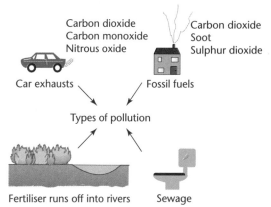

Carbon dioxide
Carbon monoxide
Nitrous oxide

Car exhausts

Carbon dioxide
Soot
Sulphur dioxide

Fossil fuels

Types of pollution

Fertiliser runs off into rivers

Sewage

Acid rain

Fig. 14.3

Acid rain is caused when the impurity sulphur, which is found in fossil fuels, is burned.

> **KEY POINT**
>
> **Sulphur burns with oxygen to form sulphur dioxide.**
> $$S + O_2 \rightarrow SO_2$$

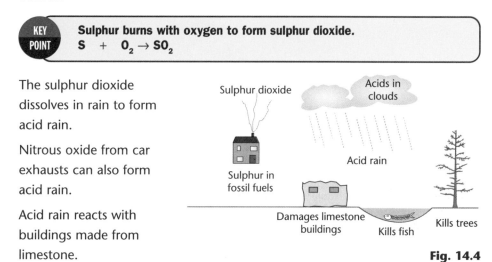

The sulphur dioxide dissolves in rain to form acid rain.

Nitrous oxide from car exhausts can also form acid rain.

Acid rain reacts with buildings made from limestone.

Sulphur dioxide

Acids in clouds

Acid rain

Sulphur in fossil fuels

Damages limestone buildings

Kills fish

Kills trees

Fig. 14.4

It kills trees and turns lakes acidic, thus killing the fish.

Environmental Impact Statement

An **Environmental Impact Statement** must be produced whenever a major new development is undertaken.

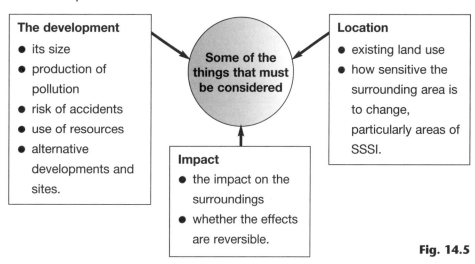

The development
- its size
- production of pollution
- risk of accidents
- use of resources
- alternative developments and sites.

Some of the things that must be considered

Location
- existing land use
- how sensitive the surrounding area is to change, particularly areas of SSSI.

Impact
- the impact on the surroundings
- whether the effects are reversible.

Fig. 14.5

The Environmental Impact Statement can then be considered when decisions are being taken about whether the new development should go ahead.

Conflict of interests

Whenever a major new development is planned, there will always be a conflict of interest between different groups of people.

Some people, such as the developers, will want the project to go ahead. They probably expect to make a large amount of money. Other people, such as local residents, may not want the development to go ahead. They probably expect the inconvenience of disruption, noise and pollution.

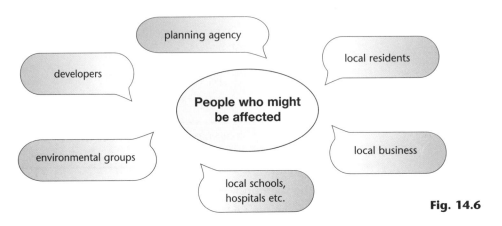

Fig. 14.6

> **KEY POINT** **Work on the development cannot begin until planning permission has been granted.**

This is a good example of Citizenship and Ideas and Evidence.

Planning permission is not granted until everyone has had a chance to register their objections. Sometimes meetings are held where people can say what they think about the development. If people cannot agree, eventually a Government Minister has to make the final decision. This process can take a long time.

PROGRESS CHECK

1. Explain what is meant by the word 'ecosystem'.
2. Explain what is meant by the word 'diversity'.
3. England used to be covered mainly by broad-leaf woodland. Explain what would happen if humans stopped controlling the environment.
4. State whether you would expect greater or lesser diversity in a non-polluted environment.
5. State two reasons why species should be conserved.
6. What is an SSSI?
7. State four main causes of pollution.
8. What gas is produced when sulphur burns.
9. State two effects of acid rain.
10. What is an Environmental Impact Statement?
11. State three groups that might be affected by a new development.

1. The environment and the organisms living in it; 2. The range of species living in an environment; 3. It would turn back to broad-leaf woodland; 4. Greater diversity; 5. Save the genes and maintain the ecosystem; 6. Site of Special Scientific Interest; 7. Burning fossil fuels, car exhausts, sewage, fertiliser run-off; 8. Sulphur dioxide; 9. Kills trees, fish and damages limestone buildings; 10. A statement that shows how a new development will affect the environment; 11. Developers, local residents, local business.

Sample GCSE question

1. Roger went on a Field Trip. He identified and counted the types of organisms in a fresh water stream. He completed the following table.

Animal name	Score	✓ if found
Shrimp	6	✓
Caddis	7	✓
Pond skater	5	
Water boatman	5	
Water beetle	5	✓
Midge larvae	2	✓
TOTAL		

This is calculating a simple biotic index.

(a) (i) Calculate the total score. [1]

20 ✓

(ii) State how many different species Roger found. [1]

4 ✓

(iii) Roger then calculated the Species Diversity Index.

Explain what this would tell Roger about the stream. [3]

The greater number of species, ✓ the higher the diversity, ✓ the less polluted is the stream ✓.

It would be easy to answer this question by saying "the more species ✓, the less pollution ✓", but this would only get two marks. This is why it is important to look to see how many marks the question is worth.

(b) Roger tested the acidity of the river and found it acid at pH6.

He thought it might be caused by acid rain.

(i) State the name of one chemical that causes acid rain. [1]

Sulphur dioxide ✓.

(ii) Explain where this chemical may have come from. [1]

Burning fossil fuels that contain sulphur ✓.

(c) The teacher told Roger that plans existed to build a factory close to the river.

(i) State two groups that might be affected if the factory was built. [2]

Local residents ✓ and developers ✓.

(ii) An Environmental Impact Statement had been produced for the new factory. Explain the purpose of the statement. [3]

To inform ✓ people about the impact ✓ of the development so that a decision ✓ could be made.

Ideas and Evidence type of question. It also includes aspects of Citizenship that you will consider in many other subject areas.

Exam practice question

1.   Joseph is a volunteer in a conservation group.

(a)   He is working to clear weeds from a pond in the local park.

(i)   Explain why the pond is an example of an artificial ecosystem.   [2]

(ii)   Give one example of a natural ecosystem.   [1]

(b)   When he has cleared the pond, he looks to see what different species he can find.

This is a chart of his results.

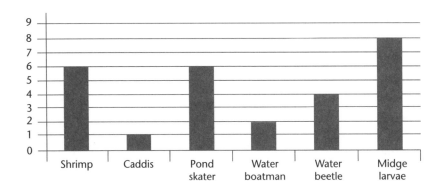

(i)   State how many different species Joseph found.   [1]

(ii)   Suggest how Joseph might use the information in the chart to calculate the 'Species Diversity Index'.   [3]

(iii) Joseph compares his results with those obtained a year ago. He finds that there are more species present now.

Explain what this tells Joseph about the condition of the pond.   [2]

(c)   The park is close to a farmer's field. The farmer uses fertilisers on his field to improve crop yield. Some of the fertiliser is washed by rain water, into the pond. The pond water turns green in colour.

(i)   State the name of this process.   [1]

(ii)   Explain why the pond water turned green.   [1]

(iii) Name one other common pollutant that can also cause this effect.   [1]

(iv) Explain why when Joseph returned to the park, he found very few fish in the pond.   [3]

(d)   Acid rain can also be a pollutant in ponds and rivers.

Explain what produces the acid rain and how it can get into rivers and ponds.   [4]

Exam practice answers

Chapter 1 Cell structure and division

1 (a) (i) Cytoplasm, chloroplast, vacuole and cell wall correctly
labelled **[4]**
(ii) The leaf
(iii) Any two from: chloroplast, cell wall, vacuole **[2]**

2 (a) Minerals are taken up by active transport **[1]** this requires energy from
respiration **[1]** respiration requires oxygen **[1]**
(b) Increased oxygen in the soil **[1]** therefore more mineral uptake **[1]** so plants
grow faster **[1]**

Chapter 2 Humans as organisms

1 (a) 1250 **[2]**
1 mark if working out shown but answer wrong
(b) (i) Same as body temperature **[1]**
(ii) Protein **[1]**
(iii) Molecules too large to pass through the membrane **[1]**
(c) Red blood cells carry oxygen from the lungs to the
tissues **[1]** white blood cells protect us from disease **[1]**
(d) (i) Kidney machines do not operate for twenty four hours
a day blood has to be returned from the machine at body temperature blood
has to be prevented from clotting while in the machine cross contamination
has to be prevented **[3]**
(ii) Answers will vary but should show some understanding of the issues
involved. **[3]**

2 (a) (i) Lungs **[1]**
(ii) (See diagram below) **[1]**
(iii) (See diagram below) **[1]**

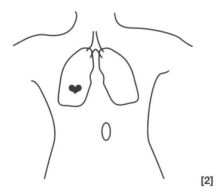

[2]

(b) Oxygen **[1]** carbon dioxide **[1]**
(c) (i) Anaerobic **[1]**
(ii) Lactic acid **[1]**
(d) Diaphragm contracts and lowers **[1]** while the intercostal muscles contract and
raise the ribs.**[1]** This increases the volume in the lungs and reduces pressure so
air is pushed in from outside **[1]**
(e) Oxygen molecules must be right size **[1]** fewer in number than those in the lungs
[1]

Exam practice answers

Chapter 3 Green plants as organisms

1 (a) Photosynthesis [1]
   (b) Carbon dioxide [1] from the air [1]
   (c) (i) Stomata (hole at bottom of leaf) [1]
       (ii) Chloroplasts inside cells near upper surface [1]
       (iii) To trap as much sunlight energy as possible [1]
   (d) Because there is no sunlight at night, [1] plants will respire and use oxygen [1]
   (e) Plants need light [1] for photosynthesis [1]
   (f) The rate photosynthesis increases [1] as light levels increase. [1]
       However eventually there is not enough carbon dioxide to make the rate go any faster [1]

2 (a) Phototropism [1]
   (b) (i) Plant will bend towards the light [1]
       (ii) Auxin makes cells get longer. [1] As auxin is on the left side, only those cells will elongate causing the shoot to bend [1]
   (c) (i) Water molecules will have more kinetic energy, [1] therefore moving faster out of the stomata and away from the plant [1] This will maintain a high diffusion gradient [1] causing a faster transpiration rate. [1]
       (ii) Wind [1] humidity [1]
   (d) When plants have plenty of water the cells are turgid [1] and press against the cell wall making the cells hard. [1] When they have too little water the cells do not press against the cell wall and they are soft, making the plant wilt. [1]
   (Reference to pumping air into a bicycle tyre as an example)

3 (a) (i) Chlorophyll – yellow leaves [2]
       (ii) Making protein – stunted growth [2]
   (b) Fertiliser makes algae grow – water turns green – blocks of light from lower algae who die – bacteria rot dead algae – bacteria use up oxygen – fish die through lack of oxygen

                                                                                    [6]
   (c) Answers will vary but consideration must be given to conflicts of interest between the farmer, the environmentalist and the effect on the ecosystem [4]

Chapter 4 Variation, inheritance and evolution

1 (a) Inherited – height weight etc. [1] environmental – sun tan or dyed hair [1]
   (b) She had different DNA [1] whereas her twin sisters had identical DNA because they were formed from the same fertilised ovum [1]
   (c) (i) CC Cc [1] Cc cc [1]
       (ii) Z [1]
       (iii) 25% [1]
       (iv) Carrier [1]
       (v) CC [1]
       (vi) CC or Cc [2]
   (d) Answers will vary, but some understanding of the issues involved should be shown. [3]

2 (a) (i) The bacteria will not be killed by that particular antibiotic. [1]
       (ii) If some bacteria have a slight resistance to one antibiotic they will not survive and pass on that resistance [1] because they will be killed by the other antibiotic [1]
   (b) Susceptible bacteria will be killed by the antibiotic. [1]
       A small number of resistant bacteria will survive. [1] These will multiply until all bacteria are resistant. [1]
   (c) Because all sexually produced offspring are different, [1] some are better adapted to survive than others. [1] These will survive and reproduce more offspring [1] while the ones that are less well adapted will not survive as long and have fewer offspring [1]

3 (a) G C T **[3]**
    (i)   Mutation **[1]**
    (ii) One amino acid will be incorrectly coded for. **[1]** This will alter the sequence of the amino acids **[1]** and alter the structure of the protein. **[1]** (Good candidates may well write about it having a greater effect if the change alters the shape of the protein).
    (iii) Chemical mutagens e.g. cigarette smoke or asbestos, or UV light **[2]**

Chapter 5 Living things in their environment

1 (a) (i) $\frac{125}{3050} \times 100$ **[1]**
          = 4.10 % **[1]**
    (ii) Energy is lost from the cow by excretion **[1]**
         egestion **[1]**
  (b) (i) Eating the beef involves the energy being passed through two organisms rather than one **[1]** therefore there are two energy transfers **[1]** so more energy is lost **[1]**
    (ii) One reason from: prefer the taste, wider range of essential amino acids **[1]**
  (c) One from: keeping the cow in warm conditions, reducing the need for the cow to search for food **[1]**

2 (a) The fat acts as a store of energy as food is sparse **[1]** the fat can be respired to produce water **[1]**
  (b) Deep roots can draw on underground water reserves **[1]** shallow roots can absorb rainwater before it evaporates **[1]** spreading out over a long distance increases the area over which water can be absorbed **[1]**
  (c) Large size and small ears gives them a smaller surface area : volume ratio **[1]** therefore helping to retain heat **[1]**
  (d) This prevents competition for food between the larvae and the adults **[1]**

Chapter 6 Classification

1 (a) A = plant
     B = plant
     C = fungi
     D = plant
     E = protoctista **[5]**
  (b) Feeds on dead **[1]** organic material **[1]** releasing enzymes onto the food **[1]**
  (c) Animals **[1]** bacteria **[1]**

2 (a) Uses a single characteristic the classify organisms **[1]**
  (b) (i) Gives organisms two Latin names **[1]** first is the genus second is the species **[1]**
    (ii) Before then organisms had common names **[1]** different organisms may have had the same name **[1]** one organisms may have had different names in different areas **[1]**
  (c) People were more widely travelled during Ray and Linnaeus's time **[1]** improved communication systems **[1]** improved documentation **[1]**

Chapter 7 Adaptation

1 (a) Two from: leaves reduced to spines, deep roots, wide spreading roots, swollen stems, green stems **[2]**
  (b) Leaves reduced to spines reduces surface area for water loss, deep roots can absorb water from deep underground, wide spreading roots can absorb water from a large area, swollen stems can store water when it is available, green stems contain chlorophyll to take over the role of the leaves **[2]**
  (c) Reptiles possess: water-tight skin **[1]** internal fertilisation **[1]** eggs with a water-tight shell **[1]**

Exam practice answers

2 (a)

insect	example of insect	type of food	brief feeding method
A	mosquito	blood	probosis pierces the skin and sucks blood
B	butterfly	nectar	proboscis uncoils and food is sucked up as though through a straw
C	housefly	human food, rotting organic matter	releases enzymes onto the food and sucks up soluble food

(b) Mosquito: may inject malaria parasite (plasmodium) **[1]** with a small amount of saliva that is injected **[1]**

housefly: pathogens picked up from sewage/faeces etc. **[1]** transferred to humans' food **[1]**

Chapter 8 Microbes and food

1 (a) *Methylophilus methylotrophus* **[1]**
   (b) Nitrogen needed to make amino acids **[1]** amino acids needed to make proteins **[1]**
   (c) Aerobic respiration **[1]**
   (d) Equipment is steam sterilised **[1]**
   (e) May be pathogenic **[1]** may produce unpleasant tasting substances **[1]**
   (f) *Fusarium* **[1]**

2 (a) To kill pathogenic microbes **[1]**
   (b) Higher temperatures would kill the bacteria **[1]**
   (c) Lactose sugar **[1]** lactic acid **[1]**

Chapter 9 Microbes, waste and fuel

1 (a) Remove large particles **[1]**
   (b) Aerobic bacteria **[1]** feed on the organic material **[1]** produce water and CO_2 **[1]**
   (c) To provide some bacteria for the activated sludge **[1]**
   (d) Water is chlorinated to kill microbes **[1]**
   (e) Fed on by bacteria **[1]** bacteria use up the oxygen **[1]** organisms die **[1]**

2 (a) Petrol **[1]**
   (b) Produces less pollution **[1]**
   (c) Petrol is not a renewable resource **[1]**
   (d) (i) Can produce higher yields of sugar cane or sugar beet **[1]**
      (ii) Starch could be fermented to produce alcohol **[1]** therefore plant material containing starch could be used **[1]**

Chapter 10 Microbes and disease

1 (a) Bacteria in the gravy were killed **[1]**
   (b) Bacteria could not fall into the flask from the air. **[1]** They would be trapped in the first bend **[1]**
   (c) Bacteria were washed from bend 'A' into the gravy **[1]** causing it to go bad **[1]**
   (d) Many possible answers, but one could be genetic
      eg. cystic fibrosis **[2]** and one could be environmental
      eg. skin cancer from sun light **[2]**
   (e) The AIDS virus enters the white blood cells **[1]** that would normally destroy the virus **[1]**

2 (a) To provide time for her body to produce antibodies to the disease **[1]**

(b) So that the drugs are in her blood when she gets bitten by the mosquito **[1]** thus ensuring that the parasite is killed when it enters her blood **[1]**

(c) If some bacteria have a slight resistance to the first antibiotic they will not survive and pass on that resistance **[1]** because they will be killed by the second antibiotic **[1]**

(d) By completing a course of treatment **[1]** and not prescribing antibiotics for trivial illnesses **[1]**

Chapter 11 *Genetics and genetic engineering*

1 (a) UGGCAUGACCUGU

All U's in correct place **[1]** rest of letters correct **[2]** -1 for any single error

(b) ACC GUA **[3]** 1 mark for each 2 bases correct

(c) (i)   X-rays, gamma rays, ultra violet light, chemical mutagens eg asbestos **[2]**

(ii)   Deletion and insertion **[1]**

(iii)   One amino acid would be wrongly coded **[1]** for the protein structure **[1]**

(d) Cystein **[1]**

2 (a) 'In glass' meaning done in a test tube **[1]**

(b) No effect **[1]**

(c) Whether the child will have cystic fibrosis **[1]** (better candidates will refer to whether there is one or two recessive alleles)

(d) Double stranded DNA **[1]** will not allow a complementary strand to attach itself **[1]**

(e) So that the 'complementary DNA' with the marker **[1]** will be able to attach itself to the single strand **[1]**

(f) The gene is not present **[1]**

Chapter 12 *Further physiology*

1 (a) Correct labels: ligament, fibrous capsule/synovial membrane, cartilage, synovial fluid **[4]**

(b) Ligament **[1]**

Bone **[1]**

Cartilage **[1]**

Tendon **[1]**

(c) A muscle can only contact **[1]** cannot actively expand **[1]** therefore has to be stretched by another muscle **[1]**

2 (a) (i)   Label on the cerebellum **[1]**

(ii)   Label on the medulla **[1]**

(iii)   Label on the cerebrum **[1]**

(b) Blood pressure detected by stretch receptors **[1]** information relayed to medulla **[1]** medulla alters heart rate or diameter of blood vessels **[1]**

(c) Bloodflow must be increased **[1]** muscles need additional oxygen or removal of carbon dioxide **[1]** due to increase respiration **[1]**

Chapter 13 *Food production*

1 (a) (i)   Eutrophication **[1]**

(ii)   Leaching or washed by rain from fields **[1]**

(iii)   Fertiliser causes algae to grow – algae near surface block out light – algae lower in water die due to lack of light **[3]**

(iv)   Bacteria rot algae using up oxygen – fish die due to lack of oxygen **[2]**

(b) (i)   Hoeing **[1]**

(ii)   Weeds may be more sensitive to weed-killer **[1]** because unlike the tall wheat crop, they have a flatter shape and more weed-killer falls on them. **[1]** Crop plants may have natural chemical resistance, **[1]** which might be through natural selection, or genetic engineering **[1]**

    (c) (i)   Reduce the yield **[1]**

        (ii) Crop rotation – prevents build up of fungal spores **[1]** use of chemicals – kills fungus **[1]** selective breeding – produces resistant varieties of crops **[1]** transfers resistant genes from other plants into crop plants **[1]**

2 (a) Answers will vary but should make reference to climate **[1]** and agricultural practices **[1]**

    (b) (i)   Fermentation – micro-organisms can be grown quickly in fermenters, **[1]** saving space **[1]**

        (ii) Gene technology – genetic modification to increase shelf life **[1]** and improve nutritional qualities such as extra vitamins **[1]**

       (iii) Manipulating reproduction – tissue culture to quickly **[1]** produce high quality, virus free plants **[1]**

    (c) (i)   Controls temperature **[1]** and some pests **[1]**

        (ii) Greater control over nutrients **[1]** and watering **[1]**

Chapter 14 *Further ecology*

1 (a) (i)   It is man-made **[1]** and would not last long if left alone **[1]**

        (ii) Any good example such as forest or ocean **[1]**

    (b) (i)   Six **[1]**

        (ii) Add up the total score for each species ie
6 + 1 + 6 + 2 + 4 + 8 **[1]** and divide by the number of species ie 6 **[1]** = 27/6 = 4.5 **[1]**

       (iii) Greater bio diversity **[1]** means a healthier pond **[1]**

    (c) (i)   Eutrophication **[1]**

        (ii) Increased growth of algae due to fertiliser **[1]**

       (iii) Sewage or washing powder **[1]**

    (d) Algae blocked out light so they died due to lack of photosynthesis – bacteria used up oxygen when rotting the algae – fish died due to lack of oxygen **[3]** burning fossil fuel that contain sulphur – produces sulphur dioxide – sulphur dioxide dissolves in rain water to form dilute sulphuric acid – acid rain falls and drains into ponds, lakes and rivers **[4]**

Index

numbers in italics refer to diagrams

For your notes

For your notes

For your notes